Jacob Berzelius (1779–1848), one of the world's leading scientists in the first half of the nineteenth century, dominated the field of chemistry, animated the cultural life of his native Sweden, and served for three decades as Perpetual Secretary of the Royal Swedish Academy of Sciences. Despite his immense stature, modern studies of his life and work have been few and scattered, and older studies have underestimated his significance. *Enlightenment Science in the Romantic Era* remedies the scarcity of accessible, modern treatments of Berzelius by bringing to a broad audience the results of recent scholarship and offers an enhanced assessment of his originality and influence.

This volume sets Berzelius's work within the contemporary scientific and cultural context of both Sweden and Europe and suggests lines for further study. Its nine substantive chapters are framed by two very different biographical portraits that establish Berzelius's principal commitments (to the values of the Enlightenment), identify his Romantic opponents, explore his ability to garner and exploit scientific and cultural resources, and offer insights into his complex personality. Other chapters explore in detail his conflicts with Romanticism, his views of life and matter, and the substance of his chemistry. The Berzelius who emerges here is a more accessible, though more complicated, figure than customarily portrayed, and his achievements become more original and enduring.

Enlightenment Science in the Romantic Era

Berzelius at the height of his career, in 1843, was a member of 84 foreign and most Swedish scientific academies. Oil painting by Olof Södermark; in the Royal Swedish Academy of Sciences. © Kungl. Vetenskapsakademien. From the archives of the Royal Swedish Academy of Sciences, used with permission.

Enlightenment Science in the Romantic Era

The Chemistry of Berzelius and Its Cultural Setting

Edited by

EVAN M. MELHADO
University of Illinois at Urbana-Champaign

and

TORE FRÄNGSMYR
Uppsala University

 CAMBRIDGE
UNIVERSITY PRESS

Published by the Press Syndicate of the University of Cambridge
The Pitt Building, Trumpington Street, Cambridge CB2 1RP
40 West 20th Street, New York, NY 10011-4211, USA
10 Stamford Road, Oakleigh, Victoria 3166, Australia

© Cambridge University Press 1992

First published 1992

Printed in the United States of America

Library of Congress Cataloging-in-Publication Data
Enlightenment science in the romantic era : the chemistry of
Berzelius and its cultural setting / edited by Evan M. Melhado and
Tore Frängsmyr.
p.   cm. – (Uppsala studies in history of science : v. 10)
Includes bibliographical references and index.
ISBN 0-521-41775-9
1. Berzelius, Jöns Jakob, friherre, 1779-1848.   2. Chemistry –
Sweden – History – 19th century.   I. Melhado, Evan Marc, 1946-
II. Frängsmyr, Tore, 1938-   .   III. Series: Uppsala studies in
history of science : 10.
QD22.B5E55   1992
5401.92 – dc20
[B]   91-45939
                                                CIP

A catalog record for this book is available from the British Library.

ISBN 0-521-41775-9 hardback

Enlightenment Science in the Romantic Era: The Chemistry of Berzelius and Its Cultural Setting is Volume 10 in the Uppsala Studies in History of Science, edited by Tore Frängsmyr.

CONTENTS

v

ILLUSTRATIONS

CONTRIBUTORS

CARL GUSTAF BERNHARD is Professor Emeritus of Physiology at the Karolinska Institute. He was President of the Royal Swedish Academy of Sciences from 1971 to 1973 and its Perpetual Secretary from 1973 to 1980. Bernhard has published many articles about Berzelius, as well as a book, *Through France with Berzelius: Live Scholars and Dead Volcanoes*, about his travels in France.

JOHN HEDLEY BROOKE is Senior Lecturer in History of Science at the University of Lancaster, where he has also been Senior Tutor and Principal of Bowland College. He has written extensively both on the history of organic chemistry and on the historical relations between science and religion. He is the author of *Science and Religion: Some Historical Perspectives* and since 1988 has been the editor of *The British Journal for the History of Science*.

GUNNAR ERIKSSON, Professor of History of Ideas and Learning at Uppsala University, has written on Linnaeus and the history of botany in Sweden, as well as on the Romantic science in Sweden. In *Kartläggarna* (1978), a Swedish-language monograph, he studied the relation between science and early industrialism in Sweden. A former editor of *Lychnos,* the Swedish yearbook for history of science, Eriksson is now studying Olof Rudbeck, a polyhistor and scientist in the generation before Linnaeus.

TORE FRÄNGSMYR is Research Professor in History of Science at Uppsala University. He has published several books, mostly about eighteenth- and nineteenth-century earth science. He is the editor of *Linnaeus, the Man and His Work* (1983), *Science in Sweden: The Royal Swedish Academy of Sciences, 1739–1989* (1989), and, with J. L. Heilbron and Robin E. Rider, *The Quantifying Spirit in the Eighteenth Century* (1990). Since

1990 he has served as the Secretary General of the International Union of the History and Philosophy of Science.

SVEN-ERIC LIEDMAN has written books on Israel Hwasser, the antagonist of Berzelius (1971), and on Carl Adolph Agardh, another contemporary of Berzelius (1991). He has also written on Marx and Marxism, the natural philosophy of Friedrich Engels, and the origins of economics as a university discipline in Sweden. He is Professor of History of Ideas and Learning at Göteborg University.

STEN LINDROTH, who died in 1980, was Professor of History of Ideas and Learning at Uppsala University from 1957 until his death. His works include a three-volume history of the Royal Swedish Academy of Sciences (1967) and a four-volume study of the Swedish history of science (1975–1981).

ANDERS LUNDGREN wrote his dissertation on Berzelius and the atomic theory (1979), and has since published several articles on the history of chemistry. An Associate Professor of History of Science at Uppsala University, he is currently studying nineteenth-century Swedish biochemistry and pharmacy.

EVAN M. MELHADO, Associate Professor of History and Chemistry at the University of Illinois at Urbana-Champaign, is the author of *Jacob Berzelius: The Emergence of His Chemical System* (1979) and co-editor of *Money, Power, and Health Care* (1988). His research focuses on chemistry in the late eighteenth and early nineteenth centuries and health care policy in the twentieth century.

ALAN J. ROCKE is Associate Professor of History of Technology and Science at Case Western Reserve University in Cleveland, Ohio. Rocke writes especially on the development of chemistry in nineteenth-century Germany.

HANS-WERNER SCHÜTT is Professor of History of Exact Sciences and Technology at the Technical University of Berlin. He has written extensively on the history of physics and technology, and is the author of *Emil Wohlwill, 1835–1912* (1972), *The Discovery of Isomorphism* (1984), and *Eilhard Mitscherlish, 1794–1863* (1991). He is currently investigating the historical relationship of science, technology, and general world view.

PREFACE

The chemist Jacob Berzelius (1779–1848) played a prominent role during his day. He animated the scientific scene in Sweden and served for three decades as the Perpetual Secretary of the Royal Swedish Academy of Sciences. At the same time, he was a leading figure within the international chemical community, traveling abroad a great deal and corresponding with colleagues the world over. However, within the history of science, he has not received attention commensurate with his stature. New studies are clearly needed to illuminate his achievements.

The idea for this book emerged some years ago, when Evan Melhado held an appointment as a guest researcher in the Office for History of Science at Uppsala University and began to collaborate with Tore Frängsmyr. We both found that historians interested in Berzelius were dispersed in many locations around the world, and we began to think of a way to bring them together to contribute to a collection of essays. Some of those we consulted agreed quickly, while others spent some time deliberating. Because both the editors and authors were located far from one another, communication was slow.

We are of course very gratified now that the book is at last ready. We wish to thank all the contributors for their cooperation and patience, and we want to acknowledge all the institutions that facilitated our work. Berzelius's manuscript collection is housed in the Center for History of Science at the Royal Swedish Academy of Sciences in Stockholm, where we benefited from excellent assistance. Economic support was graciously provided by The Bank of Sweden Tercentenary Foundation.

T.F. and E.M.M.

ABBREVIATIONS

Berzelius, *ÅB*	*Årsberättelse om Framstegen i Physik och Kemi till Kongl. Vetenskaps-Akademien af Jac. Berzelius, 1821–1840.* 20 vols. (Stockholm, 1822–1841); continued as *Årsberättelse om Framstegen i Kemi och Mineralogi afgiven af Jac. Berzelius, 1841–1847,* 7 vols. (Stockholm, 1841–1848); cf. Berzelius, *JB.*
Berzelius, "Cause of Proportions"	Berzelius, "Essay on the Cause of Chemical Proportions, and on Some Circumstances Relating to Them; Together with a Short and Easy Method of Expressing Them," *Annals of Philosophy,* 1813, 2:443–454; 1814, 3:51–62, 93–106, 244–257, 353–364.
Berzelius, *JB*	*Jahresbericht über die Fortschritte der physischen Wissenschaften von Jacob Berzelius.* Aus dem schwedischen übersetzt von C. G. Gmelin (for vols. 1–3) und F. Wöhler (for vols. 4–20). 20 vols. (Tübingen: 1822–1841); continued as *Jahresbericht über die Fortschritte der Chemie und Mineralogie eingerichtet an die schwedische Akademie der Wissenschaften von Jacob Berzelius.* 7 vols. tr. F. Wöhler (who is not mentioned on the title pages)

(Tübingen: 1842–1848); cf. Berzelius, *ÅB*.

Bref

Berzelius, *Jac. Berzelius Bref, utgifna af Kungl. Svenska Vetenskapsakademien genom H. G. Söderbaum*, 6 vols. suppl. (Stockholm and Uppsala, 1912–1935).

K. or Kungl.

Kungliga (Royal).

KVA

Kungl. Vetenskapsakademien (Royal Swedish Academy of Sciences).

KVA, *EA*

Ekonomiska Annaler, 8 vols. (Stockholm: KVA, 1807–1808).

KVA, *Handlingar*

Kungl. Vetenskapsakademiens Handlingar (Proceedings of the Royal Swedish Academy of Sciences).

LUB

Lund Universitetsbibliotek (Lund University Library).

Söderbaum, *Levnadsteckning*

H. G. Söderbaum, *Jac. Berzelius: Levnadsteckning*, 3 vols. (Uppsala: KVA, 1929–1931).

UUB

Uppsala Universitetsbibliotek (Uppsala University Library).

INTRODUCTION

EVAN M. MELHADO AND TORE FRÄNGSMYR

In Chapter 1 of this volume, the late Sten Lindroth notes that Jacob Berzelius (1779–1848), though occupying a towering position within Swedish culture, has been admired for the most part from afar. Unlike Carl von Linné (1707–1778), the other principal giant of Swedish science, Lindroth tells us, Berzelius devoted himself to a subject sufficiently recondite to defy ready appreciation by outsiders to his fields of endeavor. We may add that it is not only Swedes for whom Berzelius has been a prominent but inaccessible icon and that the difficulty of Berzelius's science has been but one cause of his remoteness. The literature on Berzelius is indeed often technically difficult; but it is also widely scattered, and, for the non-Swedish audience, much of it is linguistically out of reach.

Moreover, though it is easy enough to list the areas in which Berzelius was most active and productive (atomic theory, chemical analysis, determination of atomic and molecular weights and formulas, chemical symbolism and nomenclature, electrochemical theory, mineralogy, crystallography, organic chemistry), it is far from easy to state succinctly just why Berzelius should have cut so prominent and imposing a figure for both his own time and for posterity. His thought was synthetic, pulling together disparate strands from his own and others' work, and there is no single achievement, no "Berzelius's law" that by its very name epitomizes his accomplishments. Indeed, as Evan Melhado remarks in his contribution to this volume (Chapter 6), traditional accounts of Berzelius, for which nineteenth-century chemists-turned-historian set the pattern, have minimized his originality and tended to identify as the only enduring outcomes of his work the empirical information he developed. What precisely Berzelius accomplished and what he stood for, for his own and subsequent generations, have escaped concise characterization.

With this volume, we have attempted to render Berzelius approachable and his contemporary significance clearer by making broadly available the results of much modern scholarship emanating from both Sweden and elsewhere, and in the process suggesting topics for further study and

1

research. By modern scholarship, we mean largely two genres of writing that transcend the nineteenth-century mold, particularly Swedish works in the historiographic tradition of Johan Nordström and studies crafted largely in the spirit of what might be called (for lack of a better term) the anglophone historiography that until recently dominated the history of science.[1] More recent methodologies, such as social history, social constructivism, discourse analysis, and those portraying science as practice (as opposed to knowledge) are not represented here (though some of the essays bear traces of their influence) because their practitioners thus far have not brought Berzelius into their purview. Existing scholarship, even if not marked by the latest methodological trends, is nevertheless of high quality, and when assembled in a volume such as this, can give a broad, if still incomplete, view of Berzelius and his place in the scientific world of his own country and of Europe. The gaps in that view also serve the purpose of suggesting topics for further investigation, particularly those that may be amenable to the newer historiographic approaches.

Lindroth's essay is a good place to begin. A towering figure of the history of science in Sweden, Sten Lindroth (1914–1980) succeeded in 1957 to the first Swedish professorial chair in the history of science (*idé-och lärdomshistoria,* more accurately translated as "history of ideas and learning"), created in 1932 for his mentor, Johan Nordström (1891–1967).[2] Despite his early training in the natural sciences, Lindroth shared with Nordström a profoundly humanistic perspective on science. That perspective gave greater breadth to the scholarly tradition they inaugurated than many of its foreign models and steered it away from the positivism, whiggishness, and internalism that marked the historiography of science pursued elsewhere. Instead, they practiced a methodologically unselfconscious immersion in the primary sources, aimed to link science, technology, and medicine with broader cultural themes, and cultivated a

1 The Nordström tradition is briefly characterized in subsequent discussion. With the term "anglophone historiography" we refer to the largely internalist or intellectual history of science that emphasizes the empirical and theoretical content of science; selects, in its treatment of external factors, primarily the intellectual ones; and analyzes controversies and theory change in science chiefly in terms of its rational and empirical elements. See Tore Frängsmyr, "Science or History: George Sarton and the Positivist Tradition in History of Science," *Lychnos,* 1973/74:104–144; and Thomas S. Kuhn, "Mathematical versus Experimental Traditions in the Development of Physical Science," *Journal of Interdisciplinary History,* 1976, 7:1–31, reprinted in Thomas S. Kuhn, *The Essential Tension: Selected Studies in Scientific Tradition and Change* (Chicago: University of Chicago Press, 1977), pp. 31–65. Though the origin of this approach cannot be traced exclusively to anglophone scholars, it became dominant among them, and their example inspired much similar work elsewhere.

2 Tore Frängsmyr, ed., *History of Science in Sweden: The Growth of a Discipline, 1932–1982.* Uppsala Studies in the History of Science, Vol. 2 (Stockholm: Almqvist & Wiksell, 1984).

style calculated to express the richness and nuances of the past.[3] The methodological debates so prevalent in both anglophone and francophone historiography thus scarcely characterize the work descending from Nordström, which is marked instead by a profound appreciation for the complexity of the past, a keen instinct for the central themes, and a developed talent (shown above all by Lindroth) to illuminate those themes by deft exploitation of rich detail. However, the Nordström tradition does share the anglophone view of science as knowledge and of controversy and theory change as matters defined chiefly by the intellectual and empirical dimensions of science. In conjoining these two traditions, this volume offers a view of Berzelius that sets him within the broad context of contemporary intellectual culture, both philosophical and scientific, and portrays the evolution of ideas and work in the light of his own and his contemporaries' intellectual and empirical commitments.

The translation of Lindroth's essay has been updated and annotated and supplied with references to subsequent scholarship that has qualified some details offered in the original. However, even without modification, as a portrait of Berzelius Lindroth's piece has stood the test of time. It remains a masterly account of Berzelius's personality and beliefs, and virtually all of the themes it develops are pursued by other authors in this collection: Berzelius's passionate engagement, his combativeness, his experiences on travels abroad, his commitment to pure research and the emphasis he placed on applied science, his Enlightenment heritage, his views about method and about the nature of matter and life, his ideas about education, and his battles with Romanticism. Lindroth has brought to life a formidable personality, struggling to uphold the values of the Enlightenment in a culture increasingly saturated with a far different spirit.

Chapters 2, 3, and 4, by Sven-Eric Liedman, Gunnar Eriksson, and Anders Lundgren, are also products of the Nordström tradition (though Lundgren's exhibits traces of other historiographic modes). Their studies and that of Alan Rocke (Chapter 5) pursue three philosophical themes raised by Lindroth, Berzelius's battles against the Romantics, his atomism, and his views of vitalism and materialism.

Liedman analyzes the Romanticism of early nineteenth-century Swedish science and Berzelius's attitude toward it by exploring Berzelius's connections with two prominent Swedish contemporaries, Carl Adolph Agardh (1785–1859) and Israel Hwasser (1790–1860). Focusing on controversy about educational reform, Liedman shows that Berzelius was not necessarily completely at odds with the Romantics. However, unquestionably

---

3 Gunnar Eriksson, "Introduction: Sten Lindroth (1914–1980)," in Sten Lindroth, *Les chemins du savoir en Suède de la fondation de l'université d'Upsal à Jacob Berzelius: Études et portraits.* Ed. and tr. Jean-François Battail. Archives internationales d'histoire des idées, Vol. 126 (Boston: Martinus Nijhoff, 1988), pp. 1–11.

his position was to carry forward the Enlightenment tradition that, while valuing pure science, anticipated practical applications from the sciences and sought to foster them in appropriate educational institutions. As for the philosophy of science, Liedman illustrates Berzelius's complete impatience with the Romantics' habit of ignoring the substantive achievements of the Enlightenment style of science and substituting a florid sort of speculation. These cases amply illustrate Lindroth's observation that Berzelius, in representing eighteenth-century values, belonged to the old school.

Eriksson explores the deep background to Berzelius's atomism and his differences with the Romantics about the structure of matter. In his essay, a revised and updated translation of a piece first published in Swedish in 1967, he identifies the two leading philosophical approaches to the nature of matter: classical atomism (couched in vague terms and linked with Newtonian physics) and dynamism (articulated by Kant as a way of interpreting matter and its properties in terms of the opposition of attractive and repulsive force). Eriksson's treatment is one of the few that touch on the implications of the dynamic tradition for chemistry. From several of its German and Danish adherents, dynamism entered Sweden, flourishing, inter alia, in the circles of Uppsala in which the young Berzelius moved. Eriksson shows that at times Berzelius displayed certain affinities for dynamism, particularly in his use of the volume theory as a way to avoid the pitfalls of Daltonian atoms and in his predilection for speculation about what remained unknown of the physical world. Nevertheless, Eriksson finds that Berzelius's position was firmly rooted in the classical tradition of atomism.

Lundgren takes a different approach to Berzelius's atomism. He finds the determinants of Berzelius's views not so much in the antecedent philosophical debate as in the traditions and requirements of chemical practice. He therefore emphasizes the instrumentalism of Berzelius's principal forebears, Torbern Bergman and Lavoisier, and the common-sense requirements of practical chemistry. From this perspective, Berzelius appears as one who could scarcely dispense with atomism, but out of respect for his own empirical findings found himself obliged to reject Dalton's theory. Instead, he sought an atomism that could both explain and accommodate the laws of proportions and the empirical results of chemical analysis. Eventually, Berzelius adopted the instrumentalist position that accepted an atomic theory as a useful, if imperfect, representation of matter largely consistent with both the needs of practice and the empirical findings.

Rocke, the first of the authors here representing the anglophone tradition, proceeds mindful of historians' difficulty in characterizing Berzelius's views about the nature of life and of the substances composing living things. He finds a resolution to their dilemma by placing Berzelius within the tradition of vital materialism, which combines elements of vitalistic and materialist thought. Like Eriksson, Rocke points back to Kant, in this case, his philosophy of biology, and to figures such as Blumenbach and Reil, who pursued Kant's ideas; and he also finds roots of Berzelius's position in such figures as Haller, as well as in a variety of French physiological and medical thinkers. Hoping to lodge himself between two poles of opinion that he regarded as equally distasteful, Berzelius established not only his approach to physiological chemistry but also his position with regard to major philosophical and religious themes. In addition, Rocke finds that Berzelius's early and persistent interest in "animal chemistry" inspired his organic chemistry, informed its central themes, and thus stamped the whole field of organic chemistry in the early nineteenth century.

If Lindroth introduces Berzelius's personality and intellectual commitments, Evan Melhado, in the first of three essays on the substance of Berzelius's science, introduces the main lines of his chemistry. In Chapter 6 Melhado rejects the traditional historiography that regards Berzelius as the consolidator of revolutionary gains achieved by his predecessors and portrays his science as completely outmoded by the 1840s. Instead, he winnows Berzelius's chemistry to establish more clearly the role tradition played in it and to identify the nature of its novelties. Taking a genetic approach, he shows that Berzelius's system may be understood as an attempt to solve a fundamental problem pursued in a variety of contexts in the eighteenth century, discriminating kinds of matter. The outcomes of Berzelius's efforts to meet this goal were not fully implicit in the work of Lavoisier and Dalton, but rested on significant departures from these antecedents. By 1819 Berzelius had placed chemistry on rigorous stoichiometric foundations, he had provided a series of laws and semiempirical rules permitting the specification of most mineral substances, and he articulated the foundations that would have to underlie any effort to distinguish species of organic matter.

The next two essays, by Hans-Werner Schütt and John Brooke, pursue themes in the mineral and organic chemistry of Berzelius. Schütt, in Chapter 7, examines an important episode in the development of Berzelius's inorganic chemistry and mineralogy: the discovery of isomorphism by Eilhard Mitscherlich (1794–1863). Isomorphs are substances having the same (or very nearly the same) crystal form, and isomorphism, as understood by Mitscherlich, is a link between community of form and analogy in composition. Stoichiometrically analogous arsenate and phosphate salts, for example, possess the same crystal form. Schütt uses this case to illustrate the nature of Berzelius's chemistry and mineralogy and to exhibit Berzelius's ability, as a figure of great prestige, to develop and disseminate knowledge about the discovery and its consequences for both these sciences. To these ends, Berzelius was able to exploit his numerous personal contacts abroad, the annual progress reports he prepared as perpetual secretary of the Royal Swedish Academy of Sciences, and the laboratory facilities in which he gave Mitscherlich himself further training and set his own students to exploring the impact of Mitscherlich's discovery. By thus using the personal and professional resources at his disposal, Berzelius could powerfully affect the development of two sciences in accordance with his own interests.

In treating organic chemistry, Brooke, in Chapter 8, pursues several themes raised by other authors here: Berzelius's underlying views about the theory and practice of organic chemistry and the nature of his differences with his successors. Brooke takes issue with prevailing views about Berzelius's attitude toward vitalism, finding him cautious about the role of a vital force and constrained by empirical observations about the similarities and differences between organic and inorganic matter. The central question was whether the phenomena of inorganic chemistry ought to be taken as inevitably paradigmatic of organic chemistry, as Berzelius believed, or whether the phenomena of organic chemistry could be the source of generalizations about inorganic chemistry, as opponents, such as Auguste Laurent (1807–1853), maintained. Brooke finds through a close analysis that Berzelius's approach was far more productive than traditional accounts have suggested, that Berzelius himself was far less rigid than he has been portrayed, and that his successors owed him more than they cared to admit (or at least were able to recognize).

Finally, in Chapter 9, another portrait of Berzelius, very different from Lindroth's, closes this volume. Carl Gustaf Bernhard is not a historian of science but one of Berzelius's successors as perpetual secretary of the Royal Swedish Academy of Sciences. Having followed Berzelius on his numerous travels, Bernhard adduces a more personal and intimate sort of detail than Lindroth, and he uses it to illuminate three themes. Berzelius's avidity for travel was unusual in a day when a trip abroad was still an onerous undertaking; if he submitted to the burdens despite problems with migraine and rheumatism, it was as much for professional goals as for pleasure. By traveling, Bernhard suggests, Berzelius could supplement his extensive correspondence as a means to establish and cultivate professional connections. Moreover, by recounting Berzelius's multifaceted activities and recording his reactions to his hosts and their culture, Bernhard succeeds in exhibiting (in a way perhaps not otherwise possible) the diversity of Berzelius's interests, his keenness as an observer, and the immense esteem with which he was regarded in European scientific circles. Finally, Bernhard portrays the not always admirable, but assuredly very human qualities of a figure typically seen as unrelentingly tough and aggressive.

The image of Berzelius and his science that emerges from these essays is surely richer and more finely grained than the customary picture. They reveal how the context provided by the scientific and philosophical culture of both Sweden and Europe informed Berzelius's life and work; they discard the persistent nineteenth-century image of Berzelius, which finds him unoriginal and inflexible and his chemistry narrowly confining; they show his views of various philosophical themes such as the nature of life and matter to be more complex than previously recognized; and they provide an enhanced image of the originality and endurance of his achievements. These innovations in the study of Berzelius emerged in a scattered and linguistically diverse professional literature that has often been as inaccessible as Berzelius himself. We hope with these essays to have made him and his science more approachable and more fully appreciated.

We also hope to suggest topics for additional research; at least three loom large. First, the usual focal point of Berzelius studies, his chemistry, can still benefit from analysis, but now from the standpoint of scientific practice. Though the numerous empirical investigations that Berzelius undertook are well known, the manner in which he prosecuted them has scarcely been explored. There is no doubt that his methods, particularly for quantitative analysis of small samples, were not well precedented among chemists (though they may have been informed by practices prevailing among mineralogists). A study of the relationship among his explicit rational goals, his empirical results, and his experimental procedures may reveal that the ends of Berzelius's research were as much informed as served by his methods and that the Berzelian stamp acquired by the science of chemistry during the first half of the nineteenth century was as much a function of practice as of theory.

Second, Berzelius's role as an influential within his own country merits attention. For 30 years, Berzelius served as perpetual secretary of the Royal Swedish Academy of Sciences and in that capacity dominated the Swedish scientific world.[4] Early in his tenure, he undertook to modernize and reorganize the institution to emphasize pure science over economics and commerce. He made its *Annual Reports*, through their German and French translations, required reading throughout the scientific world. Berzelius and the academy doubtless stood in a mutually supportive re-

4 See Wilhelm Odelberg, "Berzelius as Pepetual Secretary," in *Science in Sweden: The Royal Swedish Academy of Sciences, 1739–1989*, ed. Tore Frängsmyr (Canton, MA: Science History Publications, 1989), pp. 124–147.

lationship, Berzelius using it as his institutional base and a source of prestige and the academy itself benefiting from his reputation, vigor, and commitment. Yet no thorough study exists of his role at the academy and its importance to the man or of the man to the institution. More broadly, Berzelius's place and role in the civic and cultural life of his nation also remain to be explored.

Third, perhaps the most striking feature of Berzelius's career, the immense influence he wielded on the Continent, especially after about 1820, has nowhere been assessed (though some of the elements of this story are suggested by the essays of Schütt and Bernhard), and it calls out for study in the light of the newer methodologies. How a figure located on the northern fringe of Europe, who lacked large numbers of foreign students, could so dominate the chemistry profession of the Continent is a puzzle well worth exploring. The significance, quality, and breadth of his scientific achievements doubtless garnered him great prestige, but the mechanisms for converting prestige into authority remain to be elucidated. Berzelius delivered Olympian judgments about the science of his day through the pages of the academy's widely translated *Annual Reports;* he defined the scope, content, and method of his discipline in the *Textbook of Chemistry,* which went through numerous editions and translations; he made himself the core of a network of correspondence; he developed professional contacts and influence through his frequent travels; and, though his students were few, they often proved influential bearers of his method, vision, and style. Few figures have exerted so profound an influence over an entire discipline over a period as long as two generations. The story of this achievement should reveal much about the use of social, cultural, and material resources in the advancement of scientific careers and the growth of scientific disciplines.

Berzelius therefore need not remain a remote icon viewed from afar. This volume brings him clearly into view and suggests additional studies for which the newer methodologies seem particularly apt. Though, unfortunately, linguistic barriers may persist in inhibiting additional scholarship on the career of Berzelius, the newly established Center for the History of Science at the Royal Swedish Academy of Sciences would make a congenial and attractive setting for scholars interested in further studies.

1

Berzelius and His Time

STEN LINDROTH

Jöns Jacob Berzelius – or Jacob Berzelius as he probably ought to be called – is one of the greatest names in Swedish science; only Carl von Linné can be compared with him. Yet, somehow, he is remarkably little known, though to some extent this may be explained naturally enough. Chemistry, the science to which Berzelius devoted his life, may never become as popular as botany or natural history – it is not immediately captivating or pleasurable. Even if Berzelius briefly succeeded in making chemistry something of a fashionable science in the aristocratic circles of the capital – Crown Prince Oscar and his retinue followed his experiments attentively – this was a very temporary phenomenon and primarily a tribute to Berzelius's growing international reputation. Berzelius has always been chiefly a concern of chemists and historians of chemistry; we others respect him from afar, as a monument marking the start of enormous advances made by modern chemistry in the nineteenth century.

For Berzelius himself, too, actual scientific work in the laboratory was all that really mattered. He was the pure researcher, with no other goal. For a limited period of almost unbelievable creativity, from about 1805 to 1820, he developed, on the strength of innumerable analyses and experiments, the ideas and theories that brought chemistry into a new and triumphant era: the law of constant chemical proportions and the system of chemical symbols, the contested but deeply influential electrochemical theory, the chemical classification of minerals, and physiology as a sci-

Translated by Bernard Vowles from Sten Lindroth, "Berzelius och hans tid," Vetenskapssocieteten in Lund, Årsbok, 1964:15–40.

Address given at the annual general meeting of the Scientific Society of Lund, Saturday, 23 November 1963; published here [in Swedish in the Yearbook of the society for 1964, pp. 15–40] in somewhat expanded form and provided with a minimal number of references to the sources. [The preceding note is Lindroth's; the editors' own notes or additions to Lindroth's appear in brackets; they aim to explain or clarify items Lindroth assumed were familiar to his Swedish readers. The editors have also inserted into the text a few rubrics lacking in the original.]

Sten Lindroth

This portrait has been called "the hungry Berzelius," showing the young, poor, and ambitious student at Uppsala University. From the archives of the Royal Swedish Academy of Sciences, used with permission.

ence of chemically determined processes. With his monumental textbook of chemistry, the sixth and final volume of which appeared in 1830, Berzelius crowned his scientific lifework and spoke as the highest authority among the world's chemists. But no more than other sciences can chemistry exist by itself, in the laboratory. Its fundamental problems are, in the last resort, philosophical and epistemological; they concern the structure of matter and the phenomenon of life. Berzelius was therefore compelled time and again to declare his position with regard to the important philosophical questions of the day. He did so very willingly. If ever a scientist was pugnacious and game for a fight, it was Berzelius. He took pleasure in putting his adversaries on the spot, not only his fellow chemists, but perpetrators of folly and irresponsibility wherever he might encounter them. He assumed the role of scientific arbiter and judge.

Berzelius thus became an intellectual critic, one of the most impassioned in Sweden at a time filled with aesthetic, philosophical, and political conflict. This has, of course, always been well known. The material, or at least most of it, is scattered in different parts of H. G. Söderbaum's exhaustive biography of Berzelius,[1] and his protracted feud with Israel

1 [Söderbaum, *Levnadsteckning;* see List of Abbreviations.]

Hwasser – or rather Hwasser's with him – on the subject of medical training has been discussed in several contexts.[2] But a good deal remains to be done. In particular, we need to outline the general background to Berzelius's polemics and give a balanced picture of him as a battling scientist in the Sweden of the age of Romanticism.

Rebellious Sentiments

Whether Berzelius was innately pugnacious is not known, but that is what he became under pressure of circumstances. In their way, his formative years and early career are perhaps more remarkable than is usually realized. The young Berzelius took some hard knocks, suffering one setback after another; it was as if the authorities, those who held the reins of power in science, would not let him near. Berzelius's well-known autobiography makes no little play of this. The catalog of his misfortunes is impressive: During the tender years of childhood he ate the dry bread of the grudgingly tolerated stepchild; at the gymnasium at Linköping he was [nearly] birched for illicit use of firearms, and he was sent off to the university with a humiliating final report that has become legendary; as a medical student at Uppsala in the 1790s, he became the innocent victim of rivalry between his professors; at the Serafim Hospital he was insulted by the gruff old graybeard, Olof af Acrel;[3] companions and colleagues swindled him in economic ventures and left him to his fate; he was cheated out of a promised professorship at Stockholm; the first essays he submitted to the Royal Swedish Academy of Sciences were rejected. That Berzelius's account of the misfortunes of his youth is exaggerated and highly colored is probable, indeed in part undeniable. Close examination has revealed that generally his autobiographical notes are distinctly unreliable with regard to detail and require constant checking against other source material.[4] Nevertheless, they do give us a reliable picture of how he himself was later to judge his own rise to power and splendor and the difficulties that were put in his way. They must have seemed real enough to Berzelius, and they certainly left their mark on his character and personality.

It was not only in childhood that he was compelled to stand on his

2 [See Sven-Eric Liedman in this volume, Chapter 2.]
3 [Olof af Acrel (1717–1806) was a pioneer in Swedish surgery and surgical training. He was a founder of Serafim Hospital (which opened in 1752) and head of surgery there, and in 1776 he became general director of all hospitals in Sweden, holding the post until the age of 83.]
4 Cf. in particular the study by A. Blanck, "Berzelius som medicine studerande," *Lychnos*, 1948/49:168–205. For Berzelius's well-known final report from the gymnasium of Linköping, see R. Ekholm, *Serta lincopensis* (Linköping, 1963), pp. 83ff.

own two feet, to make his own way. The opposition or indifference of those around him hardened and spurred him; all the unrelenting will-power in his makeup was stiffened in this struggle for survival. Perhaps he had to pay a certain price – in his youth he had seen the ways of the world, and the bitter and cynical view of human nature that came to be characteristic of him must have been acquired then. But stupidity and dishonesty became additional spurs to Berzelius's ambition: they made him a resister and a rebel. Quite fearless, as a 20-year-old student he lectured his professor of anatomy, Adolph Murray, for refusing him his degree. For a good many years Berzelius stood out as the angry young man of Swedish scientific life. He played this role particularly in his deal-ings with the Academy of Sciences. After the academy affronted him by rejecting his findings, he owned that he swore never to publish a line in its *Proceedings*. Not without reason he scorned the academy, then in a state of decline, and not least after he became a member in 1808 and could see at close quarters what went on behind the scenes; it appalled him.

But it was not only Berzelius's personal experiences that fueled his rebellious sentiments. The spirit of the age, the status of the sciences in the intellectual life of Sweden at that time, filled Berzelius with a not unwarranted pessimism. The heyday of Swedish natural science was long gone. The decline had set in under Gustav III; the great scientists of the Era of Liberty departed this life and were mourned, and a new and more reckless spirit spread throughout the court and the leading circles. With Gallic and literary zeal, the natural science that had once been the na-tion's pride was ridiculed – Carl Michael Bellman did not ask whether the sun moved or the earth turned on its axis,[5] the newspaper *Stock-holmsposten* jeered the pedantry of Linnean botany. After Carl Wilhelm Scheele and Torbern Bergman died in the mid-1780s, the scientific stage was truly deserted.[6] The period around the turn of the century, when Berzelius was serving his apprenticeship, was as gloomy and unfruitful as was Swedish cultural life generally. To be sure, a few were still at work, particularly the Linnean taxonomists – Thunberg, Swartz, and others – still bent doggedly over their herbaria;[7] but it was mostly rou-

---

5 [Carl Michael Bellman (1740–1795), a poet and songwriter of immense and still endur-ing popularity.]

6 [Carl Wilhelm Scheele (1742–1786) and Torbern Bergman (1735–1784), two outstand-ing names in eighteenth-century chemistry, were the last great chemical practitioners in Sweden before Berzelius. For the state of chemistry in Sweden at the end of the eighteenth century and discussion of many of the figures mentioned here by Lindroth, see Anders Lundgren, "The New Chemistry in Sweden: The Debate That Wasn't," *Osiris* (n.s.), 1988, 4:146–168.]

7 [Karl Peter Thunberg (1743–1828) was Linné's successor as professor of medicine and

tine, and the academy had difficulty filling its *Proceedings*. Chemistry declined between Bergman and "Stone Jan," the eccentric and unproductive Johan Afzelius at Uppsala – though his assistant, the deaf but amiable Anders Gustav Ekeberg, was fully conversant with the latest chemical theories and did his best to maintain Swedish scholarship.[8]

The situation scarcely improved as the new century dawned. On the contrary, scientific research faced worsening conditions in Sweden and its reputation was diminishing. Professors and physicians may have followed what was happening abroad, but only an exceptional few showed any ambition to pursue their own research. It was an unrewarding time for anyone burning with zeal to perform great deeds in science, as Berzelius discovered. Those who gave him real support – Wilhelm Hisinger, Johan Gottlieb Gahn – were, revealingly enough, technologists and mining engineers, outside official academic circles.[9] As things stood, Berzelius was left to go it alone; the apathy and incompetence around him filled him with a dismay that at times swelled into bitterness.

The New School and the Old

However, a tangible, we might almost say worthy object for Berzelius's indignation did not present itself until rather later, in the 1810s, when German Romanticism entered the country. This mysterious and beautiful gospel took hold of the young like a revelation, and from Uppsala and Stockholm it spread across the land.[10] As long as it was confined to poetry and the fine arts, there was surely no reason for a chemist to intervene. But Romanticism had totalitarian pretensions, it came with a new way of experiencing and interpreting the world, and it sought to transform even the natural sciences. The nature philosophy propounded by

botany at Uppsala. Olof Swartz (1760–1818), another follower of Linné, was Berzelius's predecessor as perpetual secretary of the academy.]

8 [Johan Afzelius (1753–1837), in fact a competent practitioner of the Swedish tradition of mineral analysis, was overshadowed by his predecessor, Bergman, and his successor, Berzelius; Anders Gustaf Ekeberg (1767–1813), though he produced little, enjoyed a significant reputation in both Sweden and abroad; he learned the new French chemistry while traveling abroad in 1789–1790 and helped introduce it into Sweden; see Lundgren, "The New Chemistry in Sweden," pp. 160, 162, 164–165.]

9 [Johan Gottlieb Gahn (1745–1818), after serving as Bergman's assistant in the 1760s, pursued a distinguished career in the mining industry; Wilhelm Hisinger (1766–1852) was a wealthy industrialist interested in mineralogy and geology; both befriended Berzelius, collaborated with him, and patronized his career.]

10 [For a brief, general introduction to Scandinavian (and especially Swedish Romanticism), see Gunnar Eriksson, "Romanticism in Scandinavia," in *Romanticism in National Context*, eds. Roy Porter and Mikulas Teich (New York: Cambridge University Press, 1988), pp. 172–190. This article touches on many of the Romantics and the documentary sources about Swedish Romanticism to which Lindroth makes reference.]

Friedrich Wilhelm Joseph von Schelling had pointed the way: With imagination and intuition as instruments, one could penetrate the mysteries of creative nature. The metaphysical basis of existence and life became the true subject of scientific research, which proceeded with gross simplifications and an uninhibited use of symbols, analogies, and images. Biologists, chemists, physicists, and physicians were entranced by this new mysticism, which provided so many shortcuts to nature's innermost recesses; between 1809 and 1811 appeared the most influential of all the handbooks in the new study of nature, Lorenz Oken's *Lehrbuch der Naturphilosophie* [Textbook of nature philosophy].

In Sweden it was the botanists who embraced most wholeheartedly the more extreme versions of nature philosophy – first among Per Daniel Amadeus Atterbom's circle of students at Uppsala and shortly thereafter in Lund, where Elias Fries and Carl Adolph Agardh became the leading apostles of Romanticism in natural science.[11] The mysteries of organic life readily invited a visionary attitude toward nature, and several Swedish physicians began now, in the mid-1810s, to align themselves with the German doctrines: They quoted Schelling, regarded the human being in a cosmic context, and in some cases preached rapturous propaganda about animal magnetism.[12] The chemists and physicists, not numerous among the younger generation, probably found it harder to keep in step, but the crucial questions of the structure of matter and the manner of action of natural forces were to an equal extent philosophical and appealed to everyone. The essential, inescapable fact was ultimately the new intellectual climate that Romanticism created. This was not in itself inimical to natural science – on the contrary, it was invested with an almost divine sublimity – but it had to be pursued in accordance with certain rules of play, in close proximity to religion and metaphysics. Rational conclusions and exact calculations authorized no conclusions about physical reality, which was in the last resort of a spiritual nature. In *Polyfem* and other journals the young Romantics lambasted and cursed the unbounded faith of the French Enlightenment in reason and its implication for natural science: materialism, which was the devil incarnate.

In these circumstances, Berzelius's stance was plain: He had grown up in the great eighteenth-century French tradition of good sense and clarity

11 [Per Daniel Amadeus Atterbom (1790–1855) was a poet, historian, and major Swedish literary figure who championed Romanticism; Elias Fries (1794–1878), prominent in nineteenth-century botany, was inter alia an expert on fungi; cf. the work of Eriksson *(Elias Fries)* cited by Lindroth; Carl Adolph Agardh (1785–1858) and his dispute with Berzelius (described briefly by Lindroth) are treated at greater length by Sven-Eric Liedman in this volume, Chapter 2.]

12 Cf. also Gunnar Eriksson, *Elias Fries och den romantiska biologin*, Lychnos bibliotek, Vol. 20 (Uppsala, 1962), pp. 116ff.

and never abandoned it as long as he lived. To him, it was the evident basis of all scientific work.

All the evidence points to the academic environment of Uppsala in the late 1790s as having confirmed Berzelius in his adherence to the most advanced ideas of the Enlightenment. This point has been discussed in an interesting and suggestive study by Anton Blanck. As a young student in the Östgöta Nation,[13] Berzelius came into contact with several influential compatriots who were active in the radical "Junta" and the circle of Benjamin Höijer, the most important among them being Gustaf Abraham Silverstolpe.[14] It is particularly interesting that the modern, revolutionary chemistry was seen in the context of the general cultural struggle. In France, Antoine Lavoisier and his theory of combustion had recently overthrown the old phlogistic system and laid the foundations of the modern subject of chemistry; his celebrated *Traité élémentaire de chimie,* published in 1789, was still very new. It was above all in the mid-1790s that Lavoisier's ideas at last began to gain ground in Sweden, and especially in circles where Berzelius moved in Uppsala. Their leading champions were among Berzelius's own teachers at the university, Anders Ekeberg and the physician Pehr Afzelius.[15] They extolled the new antiphlogistic chemistry in the same breath as they paid tribute to the "driving force of the Enlightenment" and humanity's march to perfection. The crumbling edifice of the old chemistry was demolished; in Silverstolpe's literary journal, where they were both active, the chemical revolution was accorded its place in the great European pageant, which would end in the victory of enlightenment and reason.[16]

Berzelius's Declaration of Faith

Thus Berzelius was shaped as a chemist in the light of radical progressivism; Lavoisier and the French were his mentors. Some time passed before

13 [One of the student nations, which first developed in the seventeenth century, on the model of those at the German Lutheran universities, as associations of students from the same part of Sweden or the same cathedral school. They provide practical help, comradeship, occasionally even intellectual activities, and certainly amusement even to the present day. See Sten Lindroth, *A History of Uppsala University, 1477–1977* (Stockholm: Uppsala University, 1976), pp. 84–86, 144–145, 182, 239–240, 258.]

14 [Benjamin Höijer (1767–1812) was a philosopher who anticipated themes made prominent by Fichte, Schelling, and Hegel; Gustaf Abraham Silverstolpe (1732–1824) was a publicist, pedagogue, and historian; both fostered the introduction of German Romanticism into Sweden.]

15 [Pehr Afzelius (1760–1843), brother of Johan, was a prominent physician and professor of medicine at Uppsala, who, together with Ekeberg, introduced the new chemical nomenclature into Sweden in 1795.]

16 Blanck, "Berzelius," pp. 170ff.

he himself made any declaration of faith, but when it did come it was bold, almost challenging. Berzelius took up the question of life and the life processes, fundamental to any scientific approach to nature. He did this in some well-known sections of *Föreläsningar i djurkemien* [Lectures on animal chemistry], Berzelius's first major mature work, with which he altered the course of physiological chemistry. The first volume appeared in 1806; in its introduction the question of the nature of organic life is posed.

Berzelius goes straight to the point: Living organic bodies do not occupy a place apart; life or vital force is not an indefinite "something" that is added to inanimate matter. "Life," says Berzelius, "does not lie in some foreign essence deposited in an organic or living body; its origin must be sought in the basic forces of the primary matter." Organic nature, therefore, obeys the same chemical laws as inorganic nature; what we call vital force is merely a name for the sum of the chemical and mechanical processes in the living body. The phenomenon of life consists of a continuous interplay "between the different combinations of the body's elements." In this connection Berzelius touches on the question of the origin of life. He conceives of a spontaneous generation, possibly still in progress; for him it becomes a corollary of his basically materialistic point of view. As there is no difference in principle between the inorganic and the organic, nothing is more natural than for inanimate matter to be able spontaneously to organize itself into what we call life – for it "to pass," Berzelius says, "as a result of concurrent causes, from inorganic to organic." The most simply constructed organisms, at least, such as infusoria and molds, have probably been formed in this manner; between mold and man there is only a difference in degree.[17]

The young Berzelius had launched his manifesto. With splendid clarity and natural authority of tone, he declared himself for materialism, or what was known by that name. Only matter was real; in all cases the researcher need not be concerned with anything else. Berzelius's position was far from common among chemists and biologists. Even among those lacking sympathy with Romantic animism,[18] there was a willing reverence for the phenomenon of life, not least in France, where vitalism, the belief in an enigmatically working vital force, had a particularly strong hold in medical circles at the turn of the century. The leading name was Xavier Bichat, the brilliant but tragically short-lived anatomist. Together with his compatriot Antoine François de Fourcroy, Bichat was the most

17 Jacob Berzelius, *Föreläsningar i djurkemien*, Vol. I (Stockholm, 1806), pp. 1–4.
18 [Lindroth's term, *den romantiska allbesjälningen*, possesses no single English counterpart. Insofar as he is referring to the tendency of Romanticism to regard all of nature as animate, his words may be rendered Romantic "pantheism," "animism," or "panspiritualism."]

important of Berzelius's predecessors in animal chemistry, but in the crucial question, Berzelius comes out against him.[19] It is pointless, proclaims Berzelius, to speak of the body or its parts as having an "intrinsic vitality"; this explains nothing and indeed obstructs scientific progress, if by the term is meant anything but "an as yet unknown chemical-mechanical process."[20] And yet it was the French to whom Berzelius was the most indebted. His thinking was allied with the radical philosophy of the Enlightenment and the materialists of the salons. La Mettrie, Holbach, and others, including Diderot, when he was of such a mind, had drawn up – in notorious writings abominated for their godlessness – the materialistic conception of the universe in its aggressive modern form. Beyond the world as perceived by the senses there was nothing. No spirits or forces were needed to set matter in motion, even the human soul could be interpreted with the categories of matter alone; thought and perception were corporeal properties. Thought is secreted by the brain as bile by the liver – Pierre Jean Georges Cabanis (1757–1808) had expressly said so, only a few years before Berzelius had published his lectures on animal chemistry.

Did Berzelius also profess this radically materialistic interpretation of the life of the soul? It may appear so. It is thought, he says in the section of his lectures devoted to the physiology of the brain, that the functions of the brain derive from the action of an extraneous entity called the "soul." But, on further reflection, it will be understood that they depend on the same physical and chemical laws as the other functions of the body: Judgment, memory, thought are just as much organochemical processes as the functions of the stomach, intestines, and glands, although here chemistry "ascends . . . to a higher sphere, where our inquiring spirit can never reach her." And, Berzelius adds in a note: There are learned people who deny the autonomy of the soul and thus its immortality. "This way of reasoning," he says, "which so little accords with our hopes and our almost innate feeling of the incorruptibility of the soul, is called materialism."[21] That is all – Berzelius has not so much as a word of commentary. He pronounces the detested term but is not horrified. The text hardly allows of any far-reaching conclusions with regard to his own innermost convictions, but there is no doubting the ideological tradition to which his strivings and his sympathies belonged.

19 [Xavier Bichat (1771–1802) was a prominent member of the new Paris medical school, a founder of histology, and a partisan of the view that various bodily tissues exhibit enumerable vital properties. Antoine François de Fourcroy (1755–1809) was a younger collaborator of Lavoisier, one of the founders of the new Paris medical school, and a major figure in the development of plant and animal chemistry.]

20 Ibid., p. 175.   21 Ibid., pp. 77f.

A few years later, Berzelius took up the subject again. This time it was in the address on the "Progress and Present State of Animal Chemistry," with which he duly relinquished the presidency of the Academy of Sciences.[22] The address is based largely on previously published parts of his lectures on animal chemistry. Physiological chemistry at the time was a science bubbling with activity, not least as a result of Berzelius's own efforts. Before the members of the academy, he gives a review of the results obtained and the unimagined vistas opening ahead of them. But at the same time he has become more cautious, restrained. The emphasis has shifted: Even if there has been no radical change, in his opinion, the challenging materialistic tone is absent. More mature and experienced with the years, Berzelius now stresses the immense complexity of the processes in the living organic body. What he now calls "the unknown cause of life" lies concealed in the nervous system; its functions may be chemical in their inner nature, but so intricate is their constitution that research will never succeed in fully unraveling them. The "inconceivable," says Berzelius, unfortunately plays the leading part in animal chemistry. It is of course irrefutable that the highest flights of thought and the other functions of the brain are dependent on chemical processes; but does that mean that one day our reason will be capable of investigating itself and its own nature? "I think not," declares Berzelius.[23] The circumspection that he now displays, however, implies only a further demand for uncompromising scientific stringency. Our ignorance must never tempt us to hypothesis and seductive poesy; the term vital force is still, to Berzelius, devoid of content; it leaves us no wiser.

Battling with Romanticism

Berzelius was prepared to take up the fight against Romanticism. Schooled in the chemistry laboratory, more or less a materialist, he felt only distaste and scorn for the unbridled claims of the new philosophy. Paradoxical as it may seem, if anyone belonged to the old school, it was Berzelius; Pehr Adam Wallmark and Carl Gustaf af Leopold had no more gifted

22 [The presidential terms at the academy were quite brief, and they concluded with an address by the departing incumbent, who often used the opportunity to make statements of a programmatic nature. By contrast, the most consequential functionary of the academy was its perpetual secretary, the position that Berzelius assumed in 1818 and held until his death in 1848. See Tore Frängsmyr, "Introduction: 250 Years of Science," in *Science in Sweden: The Royal Swedish Academy of Sciences, 1739–1989*, ed. Tore Frängsmyr (Canton, MA: Science History Publications, 1989), pp. 1–22, on p. 4.]
23 Berzelius, *Översigt af djurkemiens framsteg och närvande tillstånd* (Stockholm, 1812), pp. 4ff.

ally.[24] But, whereas they, as prisoners of narrow aesthetic doctrines, clung to a doomed cause, Berzelius lived as a scientist in the revolutionary tradition. What he fervently called "simple, clear, sound reason" was put in the service of progress and the future.

It was in 1811, the year after the address to the Academy of Sciences, that Berzelius made his first open attack on the nature philosophers. Strange as it may seem, he did so in the Stockholm Romantics' own mouthpiece, the short-lived *Lyceum*. The unsuspecting Lorenzo Hammarsköld had appointed Berzelius the scientific correspondent of the journal, and Berzelius responded with an article on the latest discoveries in chemistry.[25] The episode is well known. Most of it was decorous enough, but when Berzelius took up the electrochemical theory of the Hungarian chemist Jacob Winterl, it turned into a row.[26] Winterl's work contained some ingenious suggestions, but it was packed with "the most ridiculous absurdities"; the good and the bad alike were espoused by the German nature philosophers who had made it the foundation of a general theory of polarity. As a chemical theory, it was worthless, a chaos of nonsense – but that is what happens, Berzelius insists, if one tries to construct the theory of nature from one's own reason, a priori.[27] So now the young Swedish Romantics knew where Berzelius stood. Hammarsköld, the publisher of the review, dissociated himself in alarm from Berzelius's heresy in an editorial footnote, and Hammarsköld's friend Christian Stenhammar concurred wholeheartedly: Berzelius was conceited; an experimentalist and mere practitioner such as he had no right to pronounce on such lofty matters.[28] The battle lines were drawn for all time.

Apart from this outburst, Berzelius generally remained quiet during those years, at least in public. In letters to his scientific correspondents, however, he could give vent to his contempt for the buffoonery of the nature philosophers, and in conversation, too, when he saw fit. In the summer of 1812, on the occasion of the Parliament of Örebro, he met

24 [Carl Gustaf af Leopold (1756–1829) was a major poet, author, and philosopher of his day; Pehr Adam Wallmark (1777–1858), whom Leopold patronized, was a publicist, literary figure, and librarian; both were among the leading representatives of the Enlightenment in Sweden.]

25 [Lorenzo (actually Lars) Hammarsköld (1785–1827) edited *Lyceum* (published in 1810 and 1811) and, with Johan Christoffer Askelöf (1787–1848), *Polyfem* (the Romantics' Stockholm weekly, appearing from 1809 to 1812).]

26 Jacob Joseph Winterl (1732–1809) and his bizarre theories, see J. R. Partington, *History of Chemistry*, 4 vols. (London, 1961–1970), Vol. III, pp. 599–600; and H. A. M. Snelders, "Atomismus und Dynamismus im Zeitalter der deutschen romantischen Naturphilosophie," in *Romantik in Deutschland: Ein interdisziplinäres Symposion*, ed. Richard Brinkman (Stuttgart, 1978), pp. 187–201.]

27 *Lyceum*, 1811, pp. 131ff.; Eriksson, *Elias Fries*, pp. 84f.

28 Eriksson, *Elias Fries*, p. 90.

the poet Esaias Tegnér at Falkenå in the province of Närke, and it seems, to judge from the admittedly not altogether unanimous sources, that they got on immediately:[29] Tegnér's views on modern philosophy and poetry, reports Berzelius delightedly, agree with those that he long held himself.[30] But for the moment, Berzelius was not looking for quarrels. These years were the most fruitful of his life; he was deeply immersed in his pioneering works on chemical proportions and on the electrical nature of chemical attraction, and they absorbed him utterly. But in this connection, too, Berzelius could not in the end avoid taking sides concerning the speculations of the nature philosophers. The problem can be simply and clearly formulated: Did matter consist of atoms?

Berzelius, for his part, was in no doubt about the answer. The atomic theory, renewed at the turn of the century by the original work of the Englishman John Dalton, formed the very foundation of the doctrine of constant proportions. But his link to atomism meant that Berzelius had taken sides in a delicate question: Since classical antiquity, at least in eras of strict orthodoxy, the atomic theory had had a bad reputation; its adherents were stamped as materialists and atheists. The Romantics could not possibly accept it – to them nature was essentially spiritual, not a chaos of material particles. Here the two conceptions came sharply into conflict: If Berzelius wanted to prepare the way for atomistic chemistry he would first have to deal with the Romantic view of nature. Eventually he submitted the subject to examination in his important essay on phosphoric acid, published in 1816 in Gilbert's *Annalen der Physik,* and again two years later in the third volume of his *Lärbok i kemien* (1818). There was no question of stormy polemic: Berzelius elaborates his position and shows why he has found himself compelled to regard the atomic theory as indispensable to chemistry. The Romantics, the latest speculative philosophers, he says, ridicule the atomic theory and they have also frightened others away from it. They themselves are "dynamists"; they have constructed a "dynamic system" to which physicists and chemists are supposed to accommodate themselves. It is based on the idea that nature lives by spiritual forces, that matter is a product of the striving of two opposing forces in opposite directions; if one of them overpowers the other, all the matter of the universe could be gathered at a single point. The idea does not in itself shock Berzelius; he merely shows that this appealing dynamic philosophy is worthless in the chemical laboratory, because it cannot explain experimental findings. Definite chemical proportions, the fact that in nature the elements always form chemical com-

---

29 [Esaias Tegnér (1782–1846), a professor and later bishop, was one of the most famous Swedish poets. Though influenced by Romanticism, he did not share its emotional, mystical, and speculative aspects.]

30 Eriksson, *Elias Fries,* pp. 110f.

pounds in accordance with fixed numerical ratios, can be interpreted only on the assumption that matter is not infinitely divisible but corpuscular, divided into atoms that combine with each other. With his unfailing sense of what lies within the power of science, Berzelius refrains from further claims. A theory should never go "further than the need to explain the phenomena." Perhaps the chemical atomic theory is merely a "way to conceptualize" objects and not a true description of them; but it is valuable, it works.[31]

The struggle between the atomists and the dynamists lies at the very heart of the great ideological conflict of the age; it is what divided the participants. Borne on the wave of Romanticism, the dynamic theory of nature became fashionable; the Schellingian idea of polarity was its core. During the 1810s one scarcely needed to be a full-blown Romantic to profess the new dynamic philosophy and reject the atomic theory as superficial and materialistic. Samuel Grubbe in Uppsala did so, and so did the young Petrus Dahl in Lund, in a dissertation entitled *De doctrina atomorum* (1812).[32] But on the other side Berzelius was not fighting alone; he too had partisans among his scientific colleagues. In the Medical Society of Stockholm, of which Berzelius was one of the founders, nature philosophy was not at all popular. Influential members such as Erik Carl Travenfelt and Jonas Henrik Gistrén categorically repudiated it; "our science," said Gistrén, "can never become mystical."[33] Then a representative of the exact sciences came boldly forward and let fly a fearsome volley at the profundities of nature philosophy: the mathematician and astronomer Jöns Svanberg (1771–1851), at the time professor at Uppsala. A few years older than Berzelius, Svanberg very definitely belonged to the Enlightenment tradition of scientific research. In the Academy of Sciences, where he had previously served as secretary, he relinquished the presidency in February 1814, with an address entitled "Tal, öfver begrepet af naturphilosophie och dess förhållande till experimentalphysik" [On the conception of nature philosophy and its relation to experimental physics].

Svanberg's address is interesting because it shows us a new aspect of the struggle between the two schools. All the clashing, quarreling, and recriminations could also center on a single name: that of Newton. Dur-

---

31 Berzelius, "Untersuchung über die Zusammensetzung der Phosphorsäure," *Annalen der Physik*, 1816, 54:44ff; *Lärbok i kemien*, Vol. III (Stockholm, 1818), pp. 16–22.

32 Cf. Eriksson, *Elias Fries*, pp. 84, 115.

33 F. Lennmalm, *Svenska läkäresällskapets historia 1808–1908* (Stockholm, 1938), p. 36. [Jonas Henrik Gistrén (1767–1847) and Erik Carl Travenfelt (1774–1835), along with Berzelius and five others, were founding members of the Stockholm medical society and, like Berzelius, became members of the academy (Gistrén in 1804 and Travenfelt in 1812).]

ing the eighteenth century Isaac Newton had reigned supreme in the exact sciences, but with Romanticism things took on a new complexion. Johann Wolfgang von Goethe raged against his theory of colors, and dynamic philosophers had little patience for the Newtonian world mechanism. Hurt and indignant, Svanberg defended the master's greatness. Newton and Francis Bacon were his gods; only within the fold of mathematics, by the interaction of precise mathematical theory and empirical findings, could true scientific research flourish. The upstart dynamists and anti-Newtonians did not know what they were talking about; this sect from Germany, which had gained a footing in Sweden, was the enemy of light, it bred pure wretchedness, "dreams from Bedlam." But Svanberg did not go unanswered. One of the lesser phosphorists, Carl Magnus Arrhenius, delivered a vehement riposte.[34] Newton is still the main target, but Arrhenius also gives a presentation of the mysteries of dynamic nature philosophy, and it also has its pungency. Everything in nature is a process of becoming, a play of living forces, nature is a "lower Spirit become visible." Its forces are no effete constructions, lacking physical reality. Here Arrhenius invokes our celebrated chemist, Berzelius: he was "deep," an unswerving seeker after truth, who had shown with his electrochemical theory that "a living . . . Sympathy" prevails throughout nature.[35] Berzelius must have smiled wryly if he read this. He was an atomist and a Newtonian, albeit of a different sort from his brother-in-arms, Svanberg; now the opposition had laid claim to him.

When Berzelius's long tour abroad in 1818–1819 took him through Germany, his aversion for Romanticism was intensified. The students of Tübingen dismayed him. They roamed the streets in slovenly attire, with long unkempt hair and moustaches, wearing short black frock coats, and holding tobacco pipes in their hands. Berzelius, himself always carefully dressed, blamed the spirit of the times. Phosphorism, or nature philosophy as it was known in Germany, was at fault for the affected barbarism of their outward behavior – they wanted to ape the careless dress of their vigorous German ancestors. The cause of this wretchedness, he said, was "ignorance of all that is genuine, love of poetry and fine arts, and a trusting and unthinking devotion to the ideas of those whose unintelligibility has gained them a reputation for profundity."[36] As his journey there con-

34 Jöns Svanberg, *Tal, öfver begrepet af naturphilosophie* (Stockholm, 1814), pp. 31ff. For Svanberg's address and Arrhenius's reply, see also G. Ljunggren, *Svenska vitterhetens häfder*, Vol. IV (Lund, 1890), pp. 502ff.; and Eriksson, *Elias Fries*, pp. 85ff. [The term phosphorist derived from the monthly review published by the Romantics in Uppsala, *Phosphoros* (1810–1814), a counterpart of *Polyfem*.]

35 C. M. Arrhenius, *Om den falska analytiska constructionen in mathematiken* (Stockholm, 1814), pp. 50ff.

36 Berzelius, *Reseanteckningar*, ed. H. G. Söderbaum (Stockholm, 1903), pp. 307ff.

Berzelius did not like the Romantic philosophers. From his travels in Germany in 1819 he made this drawing of a student in Tübingen. The reason for this "barbarian and shabby look" is to be found in that philosophical spirit that is called *Naturphilosophie,* he concluded. From the archives of the Royal Swedish Academy of Sciences, used with permission.

tinued, he met the esoteric and unintelligible in large numbers, even among those in the exact sciences. In Dresden he quarreled about Newton's theory of colors with the professor of chemistry, Heinrick Ficinus; and in Berlin the mineralogist Christian Weiss tried for several hours to convert him from atomism to the dynamic philosophy.[37]

The Voice of Authority

Steadfast in his contempt for Romantic lunacy, Berzelius returned home to take up his new post as the secretary of the Academy of Sciences. For a decade he had been the big name in Swedish science; now he had an official position, and he did not neglect the opportunity. In March 1821, when the academy assembled to hold what was to become its annual ceremonial meeting, he gave battle. A large and eminent audience, headed

37 Ibid., pp. 345, 347. [Heinrich David August Ficinus (1782–1857) was a physician and apothecary in Dresden who became professor of physics and chemistry at the medico-surgical academy there. Christian Samuel Weiss (1780–1856) held the chair of mineralogy at the new University of Berlin from its founding in 1810 until his death. He made major contributions to crystallography and geology.]

by Crown Prince Oscar, had gathered in the academy's ceremonial lecture hall on Stora Nygatan. In the chair was the president, Gustaf Wirsén, and the flower of learned Stockholm was present.[38] After Wirsén had spoken, Berzelius came forward and presented his report on the progress of the sciences of physics and chemistry over the past year.[39] He wished at once to sweep the doorsteps of the temple clean: Politely but firmly he declared that the academy would have nothing to do with speculative research. There are two ways, Berzelius insists, of pursuing science. One can begin with experience and from that infer the causes and from them reach the laws of nature; or one can construct science a priori from a basic principle. The latter path is dangerous – the power of human imagination, he admonishes solemnly, presumes to take the place of highest reason, then "the wax in the Icarian wings melts in flight." Error and empty words will be the sole result. And he warns: Not without foreboding had the academy noted the tendency to abstract speculation beginning to spread among the rising generation here in Sweden. The consequences were inevitable; what Berzelius here calls "genuine" knowledge, practical knowledge, must languish. It was only such knowledge that permitted steady progress and served society and civilization.

Officially, in the name of the academy, Berzelius had denounced Romantic natural science. He did so as the heir to the great generation of scientific researchers of the Era of Liberty – Celsius, Klingenstierna, Wargentin – and in the same spirit as they.[40] One thing that we can state with certainty is that he did not have the whole academy behind him. The battle had not been won in his own ranks, which included some speculative heretics: Agardh and his own pupil Pehr Lagerhjelm among the younger members, his venerable excellency and nature philosopher Fredrik Wilhelm von Ehrenheim among the older.[41]

38 [Count Gustaf Fredrik Wirsén (1779–1827) was a prominent statesman of the sort who helped uphold the status and legitimacy of the academy.]

39 [Berzelius's annual reports, which he undertook in his capacity as perpetual secretary, became best known in their German translation, and through them he delivered his Olympian judgments on the science of his day. For citations, see List of Abbreviations.]

40 [Anders Celsius (1701–1744) was professor of astronomy at Uppsala, participant in Pierre Louis Moreau de Maupertuis's expedition to Lapland to measure a meridian (undertaken to settle the controversy about the shape of the earth), and propagator of the temperature scale that bears his name. Samuel Klingenstierna (1698–1765) was from 1750 the first incumbent of the newly created chair in physics at Uppsala; he was known for the study of optical systems lacking chromatic dispersion and spherical aberration, which permitted the construction of achromatic lenses. Pehr Wilhelm Wargentin (1717–1783) studied astronomy at Uppsala under Celsius, becoming perpetual secretary of the academy in 1749 and holding the post until his death. While continuing to pursue his astronomical research, he expanded the activities of the academy and brought it into closer contact with the international scientific community.]

41 [Pehr Lagerhjelm (1787–1856) was one of several of Berzelius's students elected to the academy in 1815. In 1813 he published studies conducted under Berzelius on the com-

But by this time Berzelius could feel more confident and more author-itative than ever. Now, in the early 1820s, his reputation was at its high-est; he was feted as a prince of science on his travels in Europe. He knew his own worth, and his quiet, confident self-assurance was monumental; it radiated from his broad-shouldered, powerful figure with its incipient embonpoint. Nothing could shake him when he fought for the holy cause of science; his combativeness in no way declined with the years, it merely acquired a more authoritative expression. Berzelius's letters are full of frank and blunt judgments of individuals: "rubbish," "unutterable stu-pidity," "in his total simplicity." The comments, dropped in letters to his respected friend Johan Gottlieb Gahn, refer to some of the greatest chem-ists of the period – Davy, Vauquelin, and Hausmann – when they came out against him.[42] Berzelius was indubitably a master of the art of scorn, but he usually practiced it with a discernment that the verdicts of poster-ity have not gainsaid. When he encountered humbug or moral flippancy he gave no quarter; the well-known technologist Gustaf Magnus Schwartz and the fiscal court justice Gustaf Johan Billberg, publisher of Palms-truch's botany, were among the animosities that he particularly culti-vated in his official capacity.[43] Animal magnetism, which he studied at close quarters in Berlin, enraged him.[44]

pounds of bismuth. He pursued a career in the mining industry and produced notable studies on the exploitation of hydraulic machinery in the mines. Erik Fredrik Wilhelm von Ehrenheim (1753–1828), a baron, statesman, and diplomat, was an admirer and friend of Berzelius. He was one of several leading politicians and bureaucrats elected to the academy early in the nineteenth century. Obliged to leave the government after 1809, he became a serious amateur practitioner of science. He fell under the influence of Schelling and nature philosophy, publishing the speculative *Samlingar i allmän phy-sik,* Collections in general physics (Stockholm, 1822).]

42 [Humphry Davy (1778–1829) was a distinguished practitioner whose interests (elec-trochemistry, the composition of the nitrogen oxyacids, combining weights) often closely paralleled those of Berzelius. Louis Nicolas Vauquelin (1763–1829), long an associate and collaborator of Fourcroy, is known primarily for his numerous analyses of mineral, vegetable, and animal matter. Johann Friedrich Ludwig Hausmann (1782–1859) was a prominent Wernerian mineralogist whose criticism of Berzelius's mineralogy elicited a significant evolution in Berzelius's thinking. See Evan Melhado, Chapter 6.]

43 [Gustaf Magnus Schwartz (1783–1858) was a gifted but contradictory figure promi-nent in the development of technological education and industry in Sweden. He became infatuated with Romantic mysticism and various occult practices. Berzelius, as perpet-ual secretary, eventually succeeded in driving Schwartz out of the posts, including a lectureship in technology, that he held at the academy. Johan Wilhelm Palmstruch (1770–1811), an illustrator and engraver, began a major illustrated collection of Swedish flora under the supervision of the academy; despite the high quality of the engravings, con-troversy swirled about the descriptions. After Palmstruch's death, Gustaf Johan Billberg (1772–1844), a public official and amateur scientist, took charge of the work, but incurred heavy expenses for the academy and failed to uphold the earlier standard of the engravings. As perpetual secretary, Berzelius struggled to impose discipline on the project and its director, of whom he was no admirer.]

44 Berzelius, *Reseanteckningar,* ed. H. G. Söderbaum (Stockholm, 1903), pp. 307ff.

To Berzelius, it was always the issue itself that was important; perhaps he cannot be acquitted of an element of arrogance, but there was no trace of fussy self-assertiveness or vanity. In the thick of all the disputes and the controversies, he was wonderfully calm and collected, a granite cliff and not a storm center. He did not bother with trifles. With his cool, penetrating intellect he had no patience for distinctions, decorations, or other outer conceits; in letters to his closest confidant, Hans Gabriel Trolle-Wachtmeister, he made endless fun of his newly acquired nobility and reflected irreverently on the formulation of the charter of ennoblement.[45] Berzelius generally had a liking for outrageous and high-spirited joking, and his normal demeanor was happy and cheerful: During the long, pleasant days of work in his laboratory he used to chat away, regaling his companions with anecdotes, and he was an entertaining host in his elegant bachelor apartment in the Academy of Sciences building, where ladies were also welcome. Of course this durable man had his periods of melancholy and fatigue, and at times he suffered severely from migraine; he drew heavily on his enormous capacity, and inevitably his pace slowed with time.

The unabashed claims of the nature philosophers continued to worry him. As long as there was a trace of Schellingianism left to fight, Berzelius was ready, and the 1820s brought no improvement, whether in Germany or at home. Now and then Berzelius gave his opponents what he called "a few well deserved flicks of the whip"; the money changers had to be driven from the temple. In the autumn of 1831 there came a dreadful outburst. It occurred in the course of the memorable dispute with Carl Adolf Agardh. Never did Berzelius cut a more majestic figure than in this clash; he appeared almost as divine retribution personified – or at least that is how the hapless Agardh felt.

The occasion was the second part of Agardh's *Lärobok i botanik* [Textbook of botany], which was just then in press. As is well known, the masterly and versatile man from Lund had belonged, ever since his youth, to the Romantic camp; he admired Schelling, and as a botanist, with the assistance of speculation and poetic analogy, he penetrated the mysteries of plant life. On his own initiative he sent the advance sheets of his textbook to Berzelius for perusal and criticism; for despite their differences in temperament, they were close friends, united by shared battles in the Government Committee on Education. Berzelius's reaction, therefore, came like a thunderbolt.

In a letter the length of a dissertation, Berzelius tore Agardh's botany textbook into shreds; the sections on physiology and plant chemistry bore

---

45 [Count Trolle-Wachtmeister (1782–1871), a prominent jurist and public official, was an admirer, intimate friend, and sometime student of Berzelius.]

the full brunt. It can hardly be denied that some of his criticism was pettifogging and unreasonable, but Berzelius was genuinely upset, as he had been before at animal magnetism and at the students of Tübingen. Not only was it the Romantic oddities that irritated him. Agardh's ignorance of elementary chemistry, repeatedly displayed, his offhanded treatment of terms and facts, gave Berzelius material enough for a condemnation. He did not mince words. Here were chemical incongruities, inexcusable and fundamentally erroneous assertions, nonsense, absurdities, wooly thinking. Berzelius hammers nail after nail into the coffin – Agardh wanted to flaunt his learning, but in actual fact there were dreadful gaps in his knowledge. "The author must be versed in what he is writing about," maintains Berzelius and goes on, wagging his finger imperiously: "You must polish up your style!"

But it was the speculative elements that led him to adopt the most pontifical tones. Agardh's exposition offered poetic fancy in place of strict scientific evidence. Continually he returned to a few favorite Romantic notions: the idea of a polarity running through all of animate nature (the lung and the stomach, for example, were the "poles" of the animal) and the basic conception of a deeper affinity between plants and animals. The stomach fibers corresponded to the roots of the plant, the dew was their sweat; in nature's innermost life, all was unity, identity. To Berzelius, this sort of thing was meaningless. Agardh was parading words where ideas were lacking, he scattered verses around, he was in the clutches of "scientific somnambulism." His analogies were just plays on words with which one could dazzle like a sun but "be put out with a candle snuffer" – Berzelius wielded the candle snuffer. With imposing maxims the errant botanist is instructed in the nature of science and what it demands of its practitioners. There, says Berzelius, you must have a reason for everything, every claim must be founded on experimental evidence; one must not strive for eloquence at the expense of truth. When Agardh asserts that the sun thinks for the plant kingdom, that the sun is its soul, Berzelius growls: "Truth, my friend, is the soul of science. She does not tolerate dreams, even those of genius"; behind them folly and fanaticism lay waiting.[46]

Agardh was, naturally, deeply hurt and aggrieved, though hardly contrite; he replied to Berzelius with dignity, without humbling himself, and the quarrel between the two men was soon patched up in a manner that will always do credit to both parties, particularly the maltreated Agardh. It took place entirely on a private level and never reached the general public. Yet in its way this episode marks the culmination of Berzelius's

---

46 Berzelius to Agardh, 22 November 1831, *Bref,* IV:ii E, pp. 70ff.

lifelong fight against what he saw as the false doctrines of the time. He had never expressed himself more compellingly, more pointedly. Here, if ever, the old school opposed the new – what had been good enough for Newton, Boyle, Lavoisier, and the great natural historians of the previous century should also do for Agardh.

By this time, Berzelius had already fought for his deepest convictions on another front. This was in the Government Committee on Education, the celebrated collection of brilliant minds appointed in 1825 to reform the Swedish school and university system. Berzelius had an obvious part to play, and it has often been described. In the long run his involvement in matters of official cultural policy brought him only trouble, and it culminated in the most notable of his many controversies, the battle with Israel Hwasser about the status of the Royal Karolinska Institute.[47] Here the roles were to some extent reversed. Hwasser, the dyed-in-the-wool Romantic, launched a vigorous attack on Berzelius, without mentioning him by name to be sure, and when the latter eventually replied (without naming Hwasser – the tact on both sides was admirable), he confined himself strictly to the facts of the matter; the quarrel with Hwasser, therefore, adds little to what we already know of Berzelius's own philosophical position. When viewed in its general context in the history of ideas and knowledge, however, it holds great interest. Once again, materialism, as it was called in the heat of battle, came up against Romanticism.

This was apparent from the opening moves, during the lively discussion about the universities that took place in Sweden in the 1820s. The lines were drawn: On one side stood the conservatives, the Romantics with their citadel in Uppsala, inspired by Schleiermacher and the German belief in the university as an inviolable expression of the unity of knowledge, and on the other were the radicals and the reformers, mainly Johan Gabriel Reichert and Berzelius, who wanted to make Stockholm the center of Swedish education and possibly move the whole of Uppsala University there.[48] The antagonism between the two parties came to characterize the work of the great Committee on Education and to some extent to paralyze it. Indefatigably Berzelius fought for his demands, above all, an expansion of science instruction in the schools. He had allies on the

47 [The conflict with Israel Hwasser (1790–1860) is examined by Sven-Eric Liedman in this volume, Chapter 2.]

48 Cf. P. G. Andreen, "Till frågan om ett svenskt centraluniversitet, en diskussion på 1820-talet," in *Historiska studier tillägnade Sven Tunberg* (Uppsala, 1942). [Johan Gabriel Reichert (1784–1864), a jurist and politician, was a standard-bearer of liberalism, which regarded the capital as necessarily the central location for the cultural forces of the nation.]

committee who shared his opinion, Tegnér and Agardh in particular, but the majority was against him. At the many meetings Berzelius once again found himself quarreling with the Romantics, the visionaries. Erik Gustaf Geijer and Hans Järta were his main opponents.[49] The "obscurantists," the "illiberals," they were called in Berzelius's private correspondence; Geijer, he sighed, was the leader of the sect, a sort of parliamentary representative of the phosphorists.[50] Tired and dispirited, Berzelius gradually came to see the cause as lost. "To hell with the Committee's deliberations," he wrote in a note to Agardh, and he edified Trolle-Wachtmeister with a graphic portrayal of the opposition's reactions when a vote went against them: "Järta pales, Wingård blushes, Geijer sweats and Wallin puffs."[51]

All the while, Israel Hwasser was biding his time. Hardly was the committee's voluminous report finished, at the new year, 1829, before Hwasser published his famous broadside, "Om Carolinska Institutet" [On the Karolinska Institute], under the cryptic pseudonym E.R.U.F. (an acronym from En Röst ur Fängelset [a voice from prison]). It was, of course, directed at Berzelius and at what Hwasser considered the underhanded plans, as yet in cold storage, to transfer all medical instruction to Stockholm. Hwasser's pamphlet is justly known as quite a remarkable document. Emotional, magnificent in his way, exploiting his considerable resources to the full, Hwasser floats off into impenetrable speculations through which surely none of his contemporaries could have been able to find a way. The question of the status of the Karolinska Institute becomes an aspect of the struggle between God and the devil, between good and evil. Totally captivated by the mystical-Platonic university ideal of Romanticism, Hwasser regarded the removal of the Uppsala faculty to Stockholm as a defection from the eternal order. That eventuality would mean that medicine lost its divine majesty and declined to a simple handicraft. And this, Hwasser believed, would imply the triumph of the greatest evil of the day, materialism or empiricism, the detestable way of thinking that built its world on contingent external observation. He was thus pointing his finger at Berzelius, more than anyone else. In private, Hwasser mentioned him by name – Berzelius was the "systematic, forceful leader" of the materialists in Sweden.[52] Berzelius had at last found an

---

49 [Erik Gustaf Geijer (1783–1847) was a famous poet, historian, and philosopher who shifted in the late 1830s from conservatism to liberalism. Hans Järta (1774–1847) was a politician, bureaucrat, and writer whose political evolution was the inverse of Geijer's.]

50 Söderbaum, *Levnadsteckning*, Vol. II, p. 397.   51 Ibid., p. 453.

52 Hwasser to Immanuel Ilmoni, 1832; quoted by F. Lennmalm, *Karolinska medikokirurgiska institutets uppkomst och utvekling* (Stockholm, 1910), p. 203.

opponent worthy of him in vigor and independence, with the same capacity for righteous wrath as Berzelius's own. Two of the most powerful
spirits of the age came face to face.

Perhaps that is why he did not reply to the charges. Berzelius would
have had no trouble pouring scorn on Hwasser's pretentious absurdities
– life as the identity of becoming and being, the fluidization of matter,
the lateral relationships of beings, and much more besides – but for the
moment he had enough of quarreling and kept silent. When, several years
later, in the spring of 1837, he returned under pressure of circumstances
to the controversial question, he contented himself with a powerful, fac-
tually overwhelming argument in favor of the Karolinska Institute: Medical studies were best pursued in Stockholm. He succeeded no better this
time, and a new outcry arose. Once again Hwasser aired his profundities and his hatred of materialism, empiricism, and the foundations of
modern natural science. Science, he preached, consists in "the apprehension of the eternal verities, of what is necessary in the world of thought";
it was identical with belief in the eternal.[53] Berzelius was treated no
more gently in Lund. At their deliberations in the Senate, the former
Schellingian, now Hegelian, Ebbe Samuel Bring, wielded the battle-ax.[54]
Berzelius's proposal was, according to Bring, a frightening sign of the
times: It was based on a principle that in politics had given rise to false
liberalism; in theology, to rationalism; in the natural sciences, to atomism; and in philosophy, to a superficial "common sense" or popular
philosophy.[55]

What was said was very true. The angry Lund professor gave an excellent characterization of Berzelius and the world of ideas to which he
belonged. Faith in reason and atomism were the cornerstones of the great
scientist's life and work since his youth. Plain common sense, horse sense,
was Berzelius's forte – he possessed it in abundance and in its name he
fought the Romantic visionaries as once Johan Henrik Kellgren had fought
the Swedenborgians and numerologists.[56] All this harked back to the
eighteenth century, in spirit and letter; both Bring and Hwasser knew
that. The modern perdition to which the Romantics, Hegelians, and ide-
alists offered such impassioned resistance in the mid-nineteenth century

---

53 *Handlingar rörande väckt fråga om hela medicinska undervisningens förflyttande till
   hufvudstaden* (Uppsala, 1840), p. 137.
54 [Ebbe Samuel Bring (1785–1855) was a professor of history at Lund.]
55 Ibid., p. 74.
56 [Johan Henrik Kellgren (1751–1795) was a major poet and publicist and a leading
   figure of Swedish letters in the Gustavian era. Primarily through *Stockholmsposten*,
   published over the years 1778–1795, Kellgren propagated the values of the Enlightenment, particularly its emphasis on reason and science. His battles with the rising generation of Romantics and mystics are briefly recounted by Lindroth in *Svensk lärdomshistoria*, Vol. IV (Stockholm, 1981), pp. 177–187.]

was in fact old; the old school had become the newest one, because it was inspired by the radical liberalism of the Age of Enlightenment and its rigorous demand for truth. Throughout the Romantic epoch, Berzelius had upheld the tradition of the previous century. He was not true to it in every respect. Politics did not interest him; he was, if anything, a conservative. The scientific world view and the terms for the conduct of research were what mattered, simple empiricism and enlightened reason as the basis of human knowledge. His opponents replied that this was the materialism of the old, obsolete sort, and on their own terms, perhaps they were not wrong.

Materialist or Not? Berzelius's Mature Position

Was Berzelius truly a materialist, even in a stricter, philosophical sense? We must at last attempt to answer this question. What Berzelius had to say, over the years, about the vital force and the distinctive character of organic chemistry gives important information. There is no doubt that in his youth, in his *Lectures on Animal Chemistry,* he was very close to a materialism of an eighteenth-century sort, but also, a few years later, he showed signs of greater circumspection – the innermost principles of life will always elude the chemist. Interestingly, in his more mature years Berzelius moved farther in this direction, so far in fact as to accept vitalism and praise the miracle of life. This stance appeared in the fourth volume of the great chemistry textbook, published in 1827. The words can scarcely be misunderstood: The living body is a workshop for chemical processes, but what controls and guides them we do not know; it is a "something," he says now, which has been inserted in the inorganic substance by a "force alien to inanimate nature"; this something we call the vital force. Between organic and inorganic matter there is an uncrossable line. We can never hope, declares Berzelius, "to produce organic substances with our ingenuity" in the laboratory. But he has more to offer; he gives us a world view imbued with religious feeling. The inconceivable purposefulness of organic nature must be an expression of a "sublime intelligence"; day-by-day science has revealed more of the purposes realized in living bodies, but there the researcher must come to a halt; he can only humbly admire "the wisdom we have been unable to follow."[57] The theme reechoes a few years later in the sixth volume of the textbook (1830). The secret of life, says Berzelius, is to our knowledge the most noble work of the world's creator. He speaks of human memory – it is incomprehensible what part matter, that is, water, albu-

---

57 Berzelius, *Lärbok i kemien,* Vol. IV (Stockholm, 1827), pp. 3–14.

men, and cerebral fat, plays in its sublime functions; they can only fill us with "holy awe."[58]

Berzelius had broken all ties with materialism; now, at the peak of his career, he espoused his belief in an all-wise Providence, which revealed itself in the vital force and guided creation toward certain goals. He had to some extent become religious; in the customary manner of the period he saw divine wisdom revealed in nature. But there were difficult problems, and soon Berzelius began to submit them to renewed examination. In 1828 a sensation had occurred in the world of chemistry: Berzelius's own pupil Friedrich Wöhler had for the first time produced an organic substance (urea) synthetically, in the laboratory. This delivered a deadly blow to vitalism; no longer was a mystic vital force necessary to produce organic compounds. Admittedly, it is not clear what Wöhler's discovery meant to Berzelius personally. It ought to have come as something of a shock to him. What is evident is that he reacted with a remarkable lack of excitement when in February 1828, Wöhler, agog with enthusiasm, told him in a letter that he could make urea in the laboratory.[59] When Berzelius describes the discovery in the *Annual Reports* of the Academy of Sciences (1829), he keeps strictly to the facts and does not mention the theoretical implications of the event.[60] Wöhler himself avoided any far-reaching conclusions. In his letter to Berzelius he expresses himself very cautiously – the starting materials for the synthesis of urea were, after all, extremely organic, "a nature philosopher would say . . . the organic has not disappeared and therefore an organic substance can always be produced from it again."[61]

In any case, it appears that Berzelius gradually began to return to the antivitalist position of his youth (whether or not this development was inspired by Wöhler's discovery may for the moment be left an open question). We have no right, wrote Berzelius to Agardh in 1831, to comtemplate other than chemical forces at work in living bodies.[62] Several years later, in the *Proceedings* of the academy for 1838, Berzelius developed

58 Ibid., Vol. VI (Stockholm, 1830), pp. 25–28.
59 [An insightful treatment of this episode is provided by John H. Brooke, "Wöhler's Urea and Its Vital Force? – A Verdict from the Chemists," *Ambix*, 1968, *15*:84–114.]
60 Berzelius, *ÅB, 1829*:262f.
61 Wöhler to Berzelius, 22 February 1828, in O. Wallach, ed., *Briefwechsel zwischen J. Berzelius und F. Wöhler* (Leipzig, 1901), p. 208 [editors' translation from the German].
62 Berzelius, *Bref*, IV:ii E, pp. 71f. Agardh had said "that it is not chemical forces that act within living bodies. Many say the same as you; but how do we know that? You say: because we cannot copy the works of organic nature. That proves nothing. Evidence based on inability is not evidence." The words imply that Berzelius was not at this time ready to recognize Wöhler's synthesis as the artificial production of an organic substance; otherwise he would surely have used it as an argument against Agardh.

his position on the question in more detail. He had come full circle; the words could equally well have been written in his *Animal Chemistry* 30 years earlier: A separate vital force is an unnecessary and detrimental assumption; organic processes obey the same laws as inorganic ones.[63] By this time Berzelius himself had lifted one corner of the veil surrounding organic processes with his pioneering concept of catalysis. Catalysis was a "force" that guided a particular type of process in the living body – but Berzelius saw nothing mystical in this force, which also occurred in inorganic nature and was in principle as commonplace as electricity or magnetism.[64]

If life was not an impenetrable mystery, nature as a whole was, on a higher plane. The aging Berzelius held true to a vaguely religious, idealistic view of nature; the contemplation of nature, he proclaimed, fills us with reverence for its great creator. At times, he expressed himself in edifying phrases. They were sincerely meant; he was by no means a materialist, and when Hwasser stuck that label on him it was vulgarized beyond recognition. Fortunately there were others who were more sensitive. Among them was Elias Fries, once, along with Agardh, the foremost of the Swedish Romantic botanists and thus, from Berzelius's point of view, among the lost souls. But over the years Fries had become more prudent, more intent on empirical study of reality, whereas Berzelius, as we have seen, had moved a certain distance in the other direction; the two scientists came closer together. They were thus led to explicit declarations of respect and understanding. When Fries in the autumn of 1843 sent Berzelius a copy of his newly published *Botaniska utflygter* [Botanical excursions], which was still inspired by Romanticism, Berzelius thanked him with a few rather striking lines: Reading it had convinced him that in essential matters he shared Fries's views; he had a very real admiration for him "as a thinker."[65] And Fries expressed his opinion of the great chemist some years later, after Berzelius's death; he was concerned to defend Berzelius against ill-considered invective, to save his soul after death and from death. He was, said Fries, no materialist, with time he shied away from this "hopeless doctrine" and saw even in inanimate nature the might whose wisdom finite thought cannot grasp.[66]

So placating tones were heard from both sides, and toward the middle of the century the time was ripe for reconciliation between the old and the new schools in science; they were united on faith in the creator's omnipotence. But there was never any question of capitulation on Berzelius's side. His religious view of nature, too, had its intellectual roots

63 "Om några af dagens frågor i den organiska kemien," *KVA, Handlingar*, 1838:78ff.
64 See also H. Theorell, "Berzelius och livskraften," *Nordisk medicin*, 1958, 60:1537.
65 Eriksson, *Elias Fries*, p. 419.
66 *Svenska Akademiens Handlingar*, 1852, 25:48f.; cf. Eriksson, *Elias Fries*, p. 420.

in the natural theology of the Age of Enlightenment, and in regard to the crux of the matter – the insistence of science on truth – he fought to the last. His contempt for the dreams of the nature philosophers was as strong as ever. In the summer of 1842, he spoke at a meeting of Scandinavian scientific researchers in Stockholm. In a brilliant address the elderly Berzelius opened the meeting and spoke of the morality of science, of what research demanded of its devotees. We must put the right questions to nature and wait patiently for the answers; working hypotheses can be bridges to the truth but also "switchbacks leading to mistakes and aberrations." Our age, Berzelius concludes – and we are reminded of his showdown with Agardh – loves probabilities and seductive fancies, but let us not be dazzled by will-o'-the-wisps. And he returns to the ancient Greek legend: Be warned by Icarus! It is easy to fly on the wings of hypothesis, but the sun will melt their wax and then all is lost.[67]

Both Elias Fries and Hwasser were participants at the meeting and likely heard his words – Fries, we assume, full of enthusiastic agreement and Hwasser unrepentant, lost in the fantastic dream world that Berzelius wished to crush with the full force of his personality.

67 *Förhandlingar vid de Skandinaviske naturforskarnes tredje möte i Stockholm . . . 1842* (Stockholm, 1843), pp. 24f.

2

"Truth, the Angel of Light":
Berzelius, Agardh, and Hwasser

SVEN-ERIC LIEDMAN

Jacob Berzelius was hostile to all Romantic speculation in science and medicine. In an extreme moment he once described Friedrich Schelling, the leading philosopher in the Romantic tradition, as "the most senseless of all naturalists."[1] According to Berzelius, observation and experiment were the only acceptable sources of scientific knowledge. His attitude to any kind of the organismic or holistic mode of thought so typical of Romanticism was one of stubborn opposition.

Several important Swedish natural scientists, contemporaries of Berzelius, were influenced by Romanticism. Two are discussed here: Carl Adolph Agardh (1785–1859), "the Linnaeus of the algae," and Israel Hwasser (1790–1860), professor of medicine at Uppsala and the much-loved teacher of several generations of Swedish doctors. Berzelius, Agardh, and Hwasser played leading roles in many aspects of Swedish intellectual life over a long period up to the 1850s. Agardh was not only a botanist but also a prominent economist, a politician, and an influential contributor to the debate that was in progress on the subject of education, the school, and the university. Furthermore, he ended his days as a bishop! The versatility of Hwasser was not quite so striking, but he played a significant role as a political and cultural conservative, or even a reactionary, and, although he had no real followers, as a Romantic physician he had some influence on most of his students as a role model.

Berzelius and Agardh were friends and held each other in high esteem. Gradually the differences between their scientific ideals became painfully evident to them, but this did not mean that their friendship or their mutual respect came to an end. The relationship between Berzelius and Hwasser was of a totally different kind. Hwasser fought unceasingly and, indeed, successfully to prevent the Karolinska Institute (the school of medicine in Stockholm), whose most eminent staff member was Berzelius, from gaining equivalent status as a medical school to the old medical

1 In a letter to C. A. Agardh, 18 November 1830 (LUB).

35

facilities of Uppsala and Lund. He attacked Berzelius furiously, seeing him as the chief representative of the modern tendency to reduce the high calling of the doctor to a mere handicraft. Berzelius's answer was to treat Hwasser with businesslike, chilly contempt – he seems never to have mentioned him by name.[2]

This chapter gives a fairly detailed account of the intellectual drama in which Berzelius, Agardh, and Hwasser played their respective roles. Two topics are the focus of attention. The first concerns the educational and university ideals of the protagonists. The second is the amicable but nevertheless deeply felt exchange of views in which Berzelius and Agardh engaged on the subject of Agardh's *Lärobok i botanik* [Textbook of botany, henceforth referred to as the *Textbook*].

Berzelius needs no further introduction. The following are some introductory paragraphs on the life and works of Agardh and Hwasser.

Agardh and Hwasser

Agardh was six years Berzelius's junior, having been born in 1785, the son of a small merchant in the village of Båstad in southern Sweden. His brilliant intellect soon attracted attention when he went to the University of Lund. He was a close friend of the great poet Esaias Tegnér, but, unlike him, he specialized in natural science, particularly botany and mathematics. In 1812 he was appointed professor of botany and practical economics, a combination of subjects that would have been in keeping with the utilitarian spirit of eighteenth-century Sweden but was more unusual in the Romantic period. However, Agardh possessed a rare blend of great theoretical ability and practical talent that enabled him to put new wine into old bottles. He became an outstanding, internationally renowned botanist, specializing in algae. At the same time he was a prominent economist, by far Sweden's most original thinker in the subject in the first half of the nineteenth century. Being a member of parliament, he was active in politics, proposing a radically new and very controversial economic policy that involved a program of national borrowing intended to enable the state to modernize the Swedish economy by building railways and canals.

2 This according to H. G. Söderbaum, "Berzelius och Hwasser, ett blad i den svenska naturforskningens historia," *Hygiea*, 1923, *85;17*:707; a German version of this article was published as "Berzelius und Hwasser, ein Blatt aus der Geschichte der schwedischen Naturforschung," in Julius Ruska, ed., *Studien zur Geschichte der Chemie* (Berlin: Springer, 1927), pp. 176–186, on p. 177. We may add that Agardh and Hwasser never seemed to have taken any notice of each other, even when taking part in the same debates in the Swedish parliament and supporting at least similar points of view. The reason for this is not clear, though it is likely that the brilliant and highly articulate Agardh and the rather awkward and ponderous Hwasser had some difficulty in communicating.

Agardh also held very definite opinions in the fields of basic and higher education and of science policy. He was the most radical member of the royal commission that was set up in the 1820s to propose reforms of the Swedish school and university systems. This commission was known as the Grand Education Committee, but was also referred to as *Snillekommittén* [Committee of geniuses] because of the many intellectuals it contained. Berzelius was also a member, and on most questions he sided with Agardh. Their work together on the committee led to a close friendship.

This alliance may seem strange, as Romantic or, at least, speculative ideas always played some part in Agardh's writings. However, he was never dogmatic in his attitude, and his practical and empirical ability, much approved of by Berzelius, was always evident.

When in the mid-1830s Agardh was appointed bishop of Karlstad, an appointment that was not as incongruous in those days as it would have been today, it marked the start of a new period of his life: He was not only a hierarch but also a theologian. His political and educational commitment continued, however, and he also tried to establish himself as a practical business economist, an effort that brought him disaster. He bought shares in a factory not far from Karlstad, but the enterprise was so badly managed that Agardh was bankrupted. His reaction to this misfortune was heroic. To pay off his debts, he started, at the age of 63, to publish prodigious quantities of articles and books in many different fields. In this final period, he produced his most comprehensive study in economics, *Försök till en statsekonomisk statistik öfver Sverige* [Essay on the politicoeconomic statistics of Sweden].[3] When he died, in 1859, Agardh had come close to getting out of debt.[4]

On the whole, Hwasser's life was less eventful than that of Agardh. Born in 1790, Hwasser studied medicine at Uppsala, and in 1817 he was appointed professor of practical medicine at the University of Åbo (Turku) in Finland. Up to that point, nothing that he had written or done was in any way speculative, idealistic, or Romantic; his appointment was due to his practical ability as a physician. In Åbo, however, influenced by Romantic friends – and shocked, injured, and shaken after being assaulted by a mentally deranged student and unwisely trying to steady himself with opium – Hwasser's attitude toward life and science changed

---

3 *Försök till en statsekonomisk statistik öfver Sverige,* in collaboration with C. E. Ljungberg (Karlstad, 1852–1863).
4 Among studies of Agardh's life and work may be mentioned: C. A. Adlersparre, "Carl Adolph Agardh," in Adlersparre, *Anteckningar om bortgångne samtida,* Vol. I (Stockholm, 1859), pp. 73ff.; B. Wallerius, *Carl Adolph Agardh: romantikern-politikern: Tiden i Lund* (Kungälv, 1975); G. Eriksson et al., *Carl Adolph Agardh 1785–1859: En minnesbok* (Lund, 1985); and Sven-Eric Liedman, "Carl Adolph Agardh (Gestalter i svensk lärdomshistoria 3)," *Lychnos,* 1986:71–108 (English summary, p. 104).

completely. The first clear indication of his new style was a book entitled *Om Carolinska Institutet* [On the Karolinska Institute], which appeared in 1829.[5] A vehement attack on the institute, it marked the start of Hwasser's long campaign against it, which continued until his death, in 1860. In letters written some years earlier, Hwasser had discussed in the calmest of tones an offer of a post at the Stockholm Institute (as the Karolinska Institute was originally known) from one of its professors.

Hwasser's book appeared under the pseudonym of E.R.U.F., which stood for En Röst ur Fängelset [a voice from prison]. This nom de plume reveals that Hwasser, who, as a professor in Åbo (and from 1827 to 1829 in Helsinki), was a subject of the Russian tsar, had not yet started to develop the pronounced Russian and tsarist sympathies that were to reach their peak during the Crimean War. Politically, Hwasser moved steadily toward the right from his liberal earlier years in Åbo to a reactionary position in the 1850s.

In 1830, he obtained his chair in Uppsala. There, he was a productive professor, publishing a long series of monographs on fevers, cholera, gout, and whooping cough, all characterized by a blend of classical Hippocratic medicine, problematic classification of diseases, speculative passages, and sound practical advice. He also wrote booklets and essays on moral matters such as the problems of marriage, on writers such as Sir Walter Scott, and on political issues such as Sweden's new hegemony over Norway. Personally, he seems to have radiated an unusual blend of charm, sullenness, obstinacy, and reliability. Whereas Agardh was regularly and normally described as a genius, Hwasser never was.[6]

Universities, Grandes Écoles, and the Tasks of Science

There was one field in which Berzelius, Hwasser, and Agardh all held very strong and clearly articulated views: that of the purposes of science and of higher education. Among Swedish natural scientists or physicians, they were outstanding in this respect, and for that reason alone they deserve attention.

It might well be anticipated that these three dramatis personae would belong to two distinct camps, the practical realist Berzelius to one and the Romantics to the other. This picture would be a gross oversimplification. True, Berzelius and Hwasser were poles apart, but, as has already

5 [I. Hwasser], *Om Carolinska Institutet: Betraktelse öfver det Medicinska Uppfostringsverkets närvarande tillstånd af E.R.U.F.* [On the Karolinska Institute: reflections on the current state of the medical training institution by E.R.U.F.] (Stockholm, 1829).

6 The literature on Israel Hwasser is scanty, consisting chiefly of brief articles on various aspects of his work and just one more comprehensive monograph, namely, Sven-Eric Liedman, *Israel Hwasser*, Lychnos-Bibliotek, Vol. 27 (Stockholm, 1971).

been intimated, Agardh and Berzelius were in agreement on many important questions.

The early nineteenth century forms a dramatic period in the history of higher education. During the French Revolution, the old universities were the targets of vigorous attack. They were seen as representatives of old-fashioned clerical wisdom, far from the practical ideals of the republic. In their place, the revolutionaries favored the *grandes écoles,* such as *l'École polytechnique* and *l'École des mines,* which were modern in outlook, specialized and utilitarian, their first duty being the training of professional engineers, administrators, or physicians. Leading reformers of higher education such as A. F. Fourcroy, like Berzelius a celebrated chemist, proposed the total replacement of the old universities by *grandes écoles.* Napoleon, however, implemented a compromise, enabling the *grande école* to exist side by side with a modernized type of university, one that was deprived of its traditional corporate freedom and brought under state control. The Swedish Karolinska Institute, founded in 1810 as a medical school, was inspired by the *grandes écoles* but was also a manifestation of a tendency throughout Europe to place practical medical training in the populous capitals.[7]

The early nineteenth century, however, was also a period when a totally new type of university, the German or Berlin type, appeared, where not only instruction but also research was seen as a main task of universities and where philosophy was the central discipline, the faculty of arts and sciences being preeminent.[8] Some leading German philosophers and scholars in the idealist tradition – Immanuel Kant, J. G. Fichte, Friedrich Schleiermacher, and, above all, Wilhelm von Humboldt – played major roles in formulating the new university ideal.

Both the *grande école* and the Berlin type of university represented radically new trends in higher education, and they had a common enemy: the old-fashioned, prerevolutionary university. In fundamental respects, however, they were also opposed to each other. Whereas the *grande école* was particularlized, practical, and aimed at professional instruction, the Berlin type of university was universal and theoretical and saw the *Bildung,*[9] not the practical training, of its students as its primary task.

In Sweden, Berzelius may be seen as the most important representative

---

7 Fredrik Lennmalm *Karolinska mediko-kirurgiska institutets historia* (Stockholm, 1910), Vol. I, p. 189.

8 Cf., for example, T. Nipperdey, *Deutsche Geschichte 1800–1866* (Munich, 1983), pp. 470f.; H. Schelsky, *Einsamkeit und Freiheit: Idee und Gestalt der deutschen Universitäten und ihrer Reformen* (Gutersloh, 1971).

9 The German word *Bildung* is impossible to render with a single English word. It means not only "education" but also "culture." Cf. "Bildung," *Historisches Wörterbuch der Philosophie* (Darmstadt, 1971), Vol. I, pp. 921–937.

of the *grande école* ideal. Nobody produced more persuasive and well-reasoned arguments for the Berlin ideal than Agardh. Hwasser, of course, was an unshakable opponent of the *grande école,* and in many respects he, too, was influenced by the Berlin ideal, though his view does not seem to have been so decided as that of Agardh. The link that united Agardh and Berzelius was that each advocated, eloquently and persistently, the strengthening of the position of natural sciences in both school and university. Moreover, each fought against the traditionally strong position of classical studies. Here they found an ally in Hwasser, who once declared that he hated Latin "nearly as the Devil,"[10] its only function being to perpetuate fussiness and pedantry.

In their work together on the Grand Education Committee, Agardh and Berzelius started with optimistic hopes of bringing about a substantial improvement in the position of the natural sciences. Their opponents, who argued for the continued primacy of the classics, were certainly strong, but Agardh and Berzelius believed that they would find a kindred spirit in the poet Esaias Tegnér. An old friend of Agardh, Tegnér had formerly been a professor of Greek at Lund, but by this time was the bishop of Växjö. Agardh, who was restlessly active here as elsewhere, wrote a voluminous proposal for a total reorganization of the education system and sent it to Tegnér, hoping for his support. At the same time he wrote to Berzelius, urging him, too, to contact Tegnér, as Tegnér was the man who could give them valuable assistance.[11] However, Agardh and Berzelius were mistaken. Tegnér turned out to be more conservative in educational matters than they realized and came out against Agardh's proposal.[12] The work of the committee developed into a total fiasco from the point of view of Agardh and Berzelius. In his autobiography, Berzelius gives a vivid account of the increasing narrow-mindedness of the majority of the committee. However famous this committee may be to posterity, we have to agree with Berzelius that the results of the combined efforts of its distinguished members were very meager.[13]

Berzelius and Agardh saw themselves as fighting side by side for the cause of the natural sciences. However, there was one matter on which they were not agreed, and their divergence of views was evident to them-

---

10 In a letter to his Finnish friend Immanuel Ilmoni, 24 January 1848, Helsinki University Library.
11 Letter from Agardh to Berzelius, 11 March 1827, KVA.
12 This part of the Agardh-Tegnér correspondence has been dealt with by several different researchers. Cf. L. H. Niléhn, *Nyhumanism och medborgar-fostran,* Bibliotheca historica lundensis, Vol. 23 (Lund, 1975), pp. 113f.; and, most exhaustively, T. Nordin, *Växelundervisningens allmänna utveckling och dess utformning i Sverige till omkring 1830,* Årsböcker i svensk undervisningshistoria, Vol. 53 (Uppsala, 1974).
13 Berzelius, *Biografiska anteckningar* (1901; Stockholm, 1979), pp. 91f.

selves. This was the question of the future location of Sweden's universities. Some years later, Agardh reminded his friend in a letter of their disagreement over the suggested transfer of the old universities of Uppsala and Lund to Stockholm.[14] Naturally, Agardh, a professor in Lund, spoke up for his university, whereas Berzelius, who had had his unhappy experiences at Uppsala, argued for Stockholm.

The idea of creating a new national university in Stockholm in place of the older ones found quite a lot of support among liberals in the 1820s.[15] Compared with this, the effort to strengthen the position of the Karolinska Institute was a side issue. However, the university in the capital was still only a utopian vision, whereas the institute was already in existence as a successful medical school. The controversy surrounding this school had started with its founding in 1810 and came to an end only in 1907, when it was granted the last privilege of a full-scale medical school – the right to confer the degree of doctor of medicine.

The position of the institute had to be dealt with by the committee. However, this issue was discussed in many other contexts, too, and Hwasser's vehement attack in his *On the Karolinska Institute* was just one more contribution to a widespread debate. In this debate, Agardh held his fire; Berzelius and Hwasser were the principal protagonists. It would be pointless to follow the cut and thrust of their dispute in detail. Their positions were fixed and uncompromising: Berzelius arguing for the transfer of all medical training and medical studies to Stockholm, and Hwasser proposing the total closing of the institute. Instead, we may concentrate on the general opinions on higher education held by Berzelius, Hwasser, and Agardh.

In his conception of science, Berzelius was a representative of the Enlightenment tradition. The fundamental duty of the scientist was the pursuit of sound empirical knowledge; but normally this knowledge was useful. Berzelius could also express himself more poetically, claiming that the true researcher simply followed "Truth, the angel of light," wherever it might lead him. "Often he sees clearly the practical application that a newly discovered truth may have, but nature all around is constantly revealing new secrets to him. . . . Is it strange, then, that the greater and more productive scientist is seldom the one whose application of science most benefits human existence."[16] Evidently his opinion here is that the

14 Agardh to Berzelius, undated [May 1831], KVA.
15 Cf. P. G. Andreen, "Till frågan om ett svenskt centraluniversitet, en diskussion på 1820-talet," in *Historiska studier tillägnade Sven Tunberg* (Uppsala, 1942), pp. 474f. See also Sten Lindroth, *A History of Uppsala University, 1477–1977* (Stockholm: Uppsala University, 1976), pp. 145–157.
16 Berzelius, "Om nödvändigheten för jordbrukare och ekonomer att ha insigter i chemien" [On the necessity of a sound knowledge of chemistry to agriculturalists and econ-

true scientist does not need to concern himself with the utility of his research, the practical application being the task of lesser minds and the applicability of knowledge so general that the genius does the greatest service to humanity by seeking knowledge for its own sake.

However, Berzelius's ideal scientist was never far from the world of economic application, even if he did not necessarily allow it to influence his research. In an early letter to Agardh, he expressed his admiration for a mutual friend, the French geologist A. T. Brongniart, who could combine his outstanding work as a scientist with his duties not only as a schoolmaster but even more remarkably as the director of the thriving porcelain factory at Sèvres. "Tell this to a Swedish or German professor, and he won't believe you," he exclaimed.[17]

Agardh did not protest. He himself had many highly practical interests and preoccupations, and he admired French intellectual life just as much as Berzelius did. However, his ideal savant, and especially, his ideal institution for this savant, were far from those of Berzelius. Agardh favored the new German university ideal, which he advocated most forcefully in the debates of the 1830s on the training of clergymen. The German model was characterized by the conviction that all professional training had to be located outside or at least on the periphery of the university, the chief educational task of which was the free, universal, and undirected *Bildung* of the students.

In this debate, Agardh maintains that Swedish universities have been ruined by the frequent and unnecessary compulsory examinations of the students, which serve only the aims of external control and internal pedantry. In Germany, Agardh tells his audience in the Swedish Parliament, there is a sharp dividing line between real university instruction and the examination of future civil servants. This separation is necessary, as the demands of research and *Bildung* on the one hand and practical life on the other are totally different. This difference is particularly evident in the case of theology in the university and practical Christianity outside, the task of theology being to encourage research into every aspect of Christian life and doctrine without any consideration of the creeds of the Church, whereas these creeds form the very foundation of the clergyman's practical teaching.[18]

omists], KVA, *EA,* January 1808, 5:47f. Quoted from Söderbaum, *Levnadsteckning,* Vol. I, p. 259.
17 Berzelius to Agardh, 24 April 1821, LUB.
18 Agardh in the debates in the Clerical Estate of the Swedish Parliament in 1834, reprinted in *Handlingar rörande prestbristen i Lunds stift, samt prestbildningen vid Lunds universitet . . .* (Lund, 1836), pp. 17f. and 59f. (in a special copy, UUB-K 138a, supplemented with a collection of letters, booklets, etc., which together provide the most

In this statement, Agardh seems to have gone even farther than Berzelius in his insistence on the division between the aims of university education and those of professional training. In his addresses to the meetings of the Grand Education Committee, he was not as outspoken as in the later debates on the training of clergymen, but his ideas were fundamentally the same.[19] We may add that in one respect Agardh differed from most German adherents of this university ideal in not only stressing the importance of the university's freedom but also arguing the necessity of not trying to make practical life conform to the dictates of science.

In On the Karolinska Institute, just as in many later pamphlets, articles, statements, and letters, Hwasser generally subscribed to the same ideals in higher education as Agardh, particularly underlining the importance of the university as an organic whole in contrast to specialized grandes écoles like the Karolinska Institute. He strongly emphasized that the university should be independent of the state, its task being not to serve the state but to act as a source of inspiration to state and community alike. Like society, the university ought to be a totality; and society as a whole should be permeated with its spirit.

In his view of the relationship of professional training to university study, Hwasser was a long way from Agardh's position. To him, the old faculty of medicine, emcompassing both theoretical and practical elements, was the crowning glory of the university. Hwasser was well acquainted with the German vision of the free university, but its realization in Berlin and elsewhere did not impress him at all. In his own model, examinations are few, but theoretical insight and practical instruction go hand in hand to train a wise and skillful doctor. There were now two simultaneous trends in higher education, which are in fact opposite sides of the same coin: one toward exclusively theoretical studies and the other toward strictly specialized medical training.[20]

So far, it may seem that the views of Hwasser and Berzelius were closer than those of Berzelius and Agardh. Hwasser and Berzelius at least agreed in stressing the importance of uniting theory and practice in education, whereas Agardh asserted the necessity of their institutional separation. However, Hwasser and Berzelius had very different opinions on the essence of this practical instruction, and the main target of Hwasser's attack, a target that determined his line of reasoning, was the Karolinska Institute, for long Berzelius's place of work. On the other hand, as we

comprehensive view of the university ideologies current in Sweden in the early nineteenth century).
19 See Agardh's "Slutanförande," in Betänkande af Comitén till öfverseende af Rikets Allmänna Underwisning . . . 1828 (Stockholm, 1829), pp. 83f.
20 Hwasser, Om Carolinska Institutet, pp. 94f.

have seen, Berzelius and Agardh could join forces in striving to improve the status of natural science. The main thing was that their aims were identical; that their motives were different was less important.

Berzelius saw a happy union of the pure pursuit of empirically based knowledge and its application to practical – for example, medical, industrial – aims as the best possible state of affairs. To Agardh, the first duty of natural science was neither to find such applications, which were in fact happy coincidences, nor to seek sound, disinterested knowledge within only a limited area. Just like the humanities, the natural sciences were to advance the cultivation of humankind, the effect of the humanities in this respect normally being overrated and that of the natural sciences underrated. According to Agardh, the priority of the natural sciences was evident not only in higher education and at the gymnasium but also in the general schooling of the common people (compulsory schooling was introduced in Sweden in 1842), a subject very close to his heart. As bishop of Karlstad and responsible in this capacity for the school system in his diocese, he worked hard to improve instruction in science not only at the gymnasium level but also in the elementary school, claiming that the common people had a shorter distance to knowledge in nature than in the humanities.[21]

In view of the fundamental difference between Agardh's and Berzelius's views of natural science, it is understandable that their common front in educational matters was brittle. It did not take long for their differences to come to the surface.

Agardh's Textbook

At the end of the 1820s, Agardh went through what seems to have been a deep personal crisis. He felt that his colleagues were slandering him, his pupils abandoning him, and his old friends neglecting him. On a journey to the Mediterranean, he fell ill and had to recuperate at the spa at Karlsbad. There he made the personal acquaintance of Friedrich Schelling, the Romantic philosopher par excellence. Schelling made a profound impression on him. Some years afterward, Agardh described their meetings as "the most interesting days of my life."[22] Schelling seems to have been fascinated by Agardh's algological studies and joined him in scrutinizing his new findings through the microscope.[23] He also encouraged him to develop his own thinking on nature and life in a treatise in German.

Hitherto, Agardh had had an ambivalent attitude toward Schelling's

21 Agardh, *Försök till en statsekonomisk statistik*, Vol. I:i, pp. 168f.
22 Agardh, *Lärobok i botanik* (Malmö, 1829–1831), Vol. I, dedication.
23 Agardh, *Berättelse om en botanisk Resa till Österrike och Nordöstra Italien år 1827* (Stockholm, 1828), pp. 12f.

philosophy. Although in many respects a Romantic in his sympathies, he had once made fun of Schelling's writings when facing some of the German philosopher's most ardent Swedish followers.[24] In actual fact, his familiarity with Schelling's philosophy seems to have been rather superficial, based on little more than a reading of certain abstracts.[25]

This cursory knowledge must not be seen merely as evidence of a careless attitude to his real sources of inspiration. There had always been and there remained a deep gulf between Schelling's natural philosophy and Agardh's own approach to natural science. To Schelling, the sciences form a hierarchy dominated by philosophy. Chemistry gives us the alphabet, physics the syllables, and mathematics enables us to read nature – but only philosophy can *interpret* what we have read, he proudly declared.[26] Agardh never accepted this view of the philosopher as a supreme judge of science. In Agardh's opinion, the naturalist had to be his own philosopher, sometimes inspired by professional philosophers but never governed by them. Schelling drew scientific inferences directly from philosophical premises; Agardh never did. During his long and varied intellectual career, Agardh played the role of mathematician, botanist, practical and political economist, educator, and theologian. However, he never proceeded from his philosophy to draw conclusions in any field in which he had not himself carried out empirical research. Even in Agardh's most speculative writings, one can trace the imprints of the *other* Agardh, the Agardh with his feet firmly on the ground.

Hence, Agardh was never a Schellingian, although his sympathy for Schelling reached its peak in the years before 1830 and found its most visible expression in his dedicating the first volume of his *Textbook* to Schelling. The dedication is in German, but the rest of the book is in Swedish. The main reason for his failure to follow Schelling's suggestion of writing in German seems to have been the peculiar haste in which he completed this, his hitherto most comprehensive and voluminous work. At the same time, there is much in the Swedish text that would not have pleased Schelling. For instance, Agardh declares that Romantic natural philosophy has not succeeded in finding answers to eternal questions of life and the nature of the organism.[27] The book did appear in a German translation a few years later – but this time without the dedication to

24 Wallerius, *Carl Adolph Agardh*, pp. 14f.
25 When talking of Schelling's early writings, Agardh was referring only to certain abstracts published in *Allgemeine Literature-Zeitung in Jena;* see his "Autopsia" and "Journal" MSS in the Värmland Museum, Karlstad.
26 Schelling, *Sämmtliche Werke* (1857; Munich, 1927), Vol. I:ii, p. 6.
27 Agardh, *Lärobok*, Vol. I, p. 79. On Schelling and his followers, he said that they only "searched for nature in general, not a definite and definitive nature [*i.e.*] not for the wonderful nature that Eternal Reason has created."

Schelling. The reason for this omission is unknown, but we may guess that Schelling's reaction, or rather his lack of reaction, to the original dedication had hurt the sensitive Agardh. Be that as it may, there is no doubt that his attitude toward Schelling became decidedly more negative than it had been at the time of the short interlude in Karlsbad.[28]

Let us now turn to the main content of the two volumes of the *Text-book*. The common theme is botany, although the first part has the sub-title *Organografi* (organography, or description of the organisms) and the second one, *Wext-Biologi* (biology of plants, or plant physiology). Agardh had planned a third, concluding volume, in which he would present his own natural system; in the absence of a competent Swedish readership for this part, it was intended to be published in Latin.[29] However, his system remained uncompleted, and we can only wonder whether the reason was the lukewarm reception of the first two volumes, the constant pressure of other activities, or the start of his search for a nonacademic career.

In the opening section of the first volume, Agardh once again maintains the supreme value of natural history in general education. Its main value does not lie in its "material usefulness," either medical or economic; from this point of view it lags behind "the younger science of chemistry." No, the value of natural history is "mainly philosophical." There are certain important questions that man cannot escape and to which natural philosophy offers tentative answers.[30] We shall not scrutinize Agardh's own philosophical answers in any detail. It is impossible to assign them to a single philosophical tradition. As a metaphysician, Agardh was an eclectic, pursuing lines of thought from a wide range of traditions – from Kant, Fichte, Jean Paul, even from Aristotle.

His biological philosophy seems to be more consistent and more original. It is interesting to find that he applied the same guiding principles as when dealing with society and human life. As an economist, educator, or politician, his position may seem to be inconsistent. Ideologically, friends and foes sometimes labeled him a conservative, sometimes a liberal, or even, in a Benthamite sense, a radical. Despite these appearances, it can be argued that from biology to politics he was actually quite consistent. His position might be called one of *organic individualism*. This very con-

28 Parenthetically, we may add that the two met once more, in Berlin, probably early in 1842. Agardh, who was now a 57-year-old bishop and a studious theologian, attended some of Schelling's famous lectures on the revelation, sitting alongside such upcoming men as Friedrich Engels, Søren Kierkegaard, Mikhail Bakunin, and Jakob Burckhardt. Afterward, Agardh reported on the lectures to the priests in his diocese. In those reports he was decidedly critical of Schelling's philosophy. Cf. Liedman, "Agardh."
29 Agardh in a letter to Berzelius, 4 November 1830, KVA.
30 Agardh, *Lärobok*, Vol. I, p. 33.

cept suggests that he combined elements that are normally attributed to different philosophical and ideological traditions, as individualism is to atomism and/or liberalism, and organicism is to Romanticism and/or conservatism.

The concept of *individuality* is fundamental in his *Textbook*, as in many of Agardh's botanical writings.[31] To understand his use of it, we must follow his argument in some detail. Agardh made some effort to construct a boundary between the living and the dead and between plants and animals. However, inorganic matter is outside his scope; *pace* Berzelius's severe criticism of his tendency to speculate, we may confirm that in this respect he differs markedly from the mainstream of Romantic natural philosophers, who are given to a priori reasoning on every aspect of nature. However, Agardh makes a clear distinction among the "four realms of nature," the fluids, the minerals, the plants, and the animals. His style here is as much Aristotelian as Romantic. He argues that every natural entity is in the process of developing to a higher level. Forms that seem to be half organic/half inorganic are potentially organic.

In this argument we find a key to Agardh's whole organic philosophy. When defining an organism, one must start not from the simplest, most primitive forms but from the highest and most complicated ones. Within the more rudimentary forms, the higher ones exist as intrinsic possibilities. In the *Textbook*, Agardh declares: "All organic nature has a tendency to regularity, but in itself it is not regular."[32] Implicitly, the regularity of a crystal is typical only of inorganic matter.

Unlike Schelling and his adherents, Agardh made a sharp division between living and inanimate nature, and he differed from them in assuming a specific *vis vitalis,* or vital force. However, when he talked exclusively of organisms most of his arguments came close to the Romantic line of reasoning. For example, according to Agardh, an organism is a totality, the parts of which cannot be understood without the totality, nor the totality without the parts; and the totality expresses an *idea.* However, Agardh put the emphasis in a way that had implications for his individualistic bent. He accentuated the role of the part and maintained that the organ in the specific organism may itself be seen as an organism at a lower level. True, it is dependent on the total organism, but potentially it is itself an independent organism.

Here Agardh did in fact use the word individual rather than organism.

---

31 Among important writings preceding the *Textbook* may be mentioned *Aphorismi botanici*, pt. 5 (Lund, 1819); *Classes plantarum* (Lund, 1825); *Species algarum* (Lund, 1828), especially "Praefatio"; and his presentations in French of his biological philosophy: *Essai de réduire la physiologie à des principes fondamentaux* and *Essai sur le développement intérieur des plantes* (both Lund, 1828).

32 Agardh, *Lärobok,* Vol. I, p. 33.

Almost poetically, he compared the fur of an animal or the feathers of a bird to parasites, which are already individualized but which still have a "weak individuality." He added that there are "individualities or totalities of different degrees." Clearly, an individuality of lower degree may be subsumed into one of higher degree.[33]

Agardh's most general theses on plants and plant anatomy must be understood in the context of these ideas about organisms and individuals. His argumentation may be summarized as follows:

1. A plant – which is in itself a total organism – consists of several unfulfilled organisms, which are dependent on the plant.
2. The degree of independence of an organism is the degree of individuality.
3. There is a natural tendency for less independent or perfect individuals to develop toward greater independence or perfection. (This point must not be taken to imply that Agardh would support a pre-Darwinian theory of natural evolution. Like most Romantics, he saw evolution in nature as an ideal and not as a real process.)

In his economic, social, and political theories, Agardh used the same fundamental concepts, though with apparently greater ease. However, in these contexts, "individual" means "individual man" and "totality" means "state" or, sometimes, "society." Agardh's most outspoken and original conviction in this field is that civilization means individualization, but that this individualization can develop only within state and society. According to him, the most fully fledged, autonomous human individual can be realized only within a strong political, social, and economic unit. In other words, collectivity comes before individuality, and the freest possible individual still lives within society, serving and supporting a strong state.

This conception seems to have traits in common with both liberalism and conservatism. Agardh differed from most liberals in supporting the community and the state, but in contrast to the conservative he stressed the importance of the individual.[34]

The *Textbook* in the Agardh-Berzelius Correspondence

Although they had become close friends during the years when they met daily on the Grand Education Committee, Agardh and Berzelius were aware that they were not in agreement in every respect. However, being

33 Ibid., pp. 144f.
34 Agardh himself was aware of the unusual ideological mixture; cf. his letters to Adlersparre, quoted in Adlersparre, pp. 142f.

allies in the campaign to strengthen the position of the natural sciences in school and higher education was a cornerstone of their friendship. To both, the meager results of the committee's work was a deep disappointment. In the years after it came to an end, they vied with one another in dismissing their fruitless endeavors. Agardh compared them to a "stormy sea voyage," which "ends with a wreck"[35]; Berzelius, as always harsher and less poetic, declared that the committee has vanished from his mind.[36]

This period around 1830 brought a deep personal crisis to Agardh. Incessantly, he tried to find new activities, new ways of escape, new outlets for his restless energy. The disputes with narrow-minded colleagues in Lund infuriated him, and the idea of leaving his chair to find a lucrative parish – in those days a normal recourse for a discontented Swedish professor – recurs frequently in his correspondence. In a particularly pessimistic letter to Berzelius, he even hinted at the possibility of finding some part-time occupation in Stockholm, or even in Paris or in London. "As you know, I am addicted to drastic cures," he added.[37] However, Berzelius seemed startled by this letter and responded: "My dear Agardh, you are ill." He tried to convince his friend that his quiet, safe position at Lund is an ideal one; besides, Lund is closer than Stockholm to Europe.[38] And Agardh stayed at Lund until, some years later, he obtained a substantially better position, the bishopric in Karlstad.

His *Textbook of Botany* is the product of some very turbulent years of his life, a fact of which Berzelius was not unaware. In 1828 Agardh had published two smaller volumes in French, in which he put forward some of the fundamental ideas of the coming *Textbook,* although the tone was by no means as Romantic as it later became and the results were presented in a somewhat Gallic, quasi-exact fashion.[39] He sent the booklets to Berzelius, who seemed rather surprised and only declared that he knew too little plant physiology to form an opinion on Agardh's theories.[40]

Two years later, Agardh sent him the first volume of his Swedish-language *Textbook.* In an accompanying letter, Agardh declared that he would have sent him the text much earlier, had he only known that Berzelius was willing to give him advice and to correct his mistakes. (Apparently Berzelius had told him at a meeting in Stockholm that he would

---

35 Agardh to Berzelius, 18 June 1828, KVA.
36 Berzelius to Agardh, 27 June 1828, LUB.
37 Agardh to Berzelius, undated [1831], KVA.   38 Berzelius to Agardh, 1831, LUB.
39 For the titles, see note 31. That Agardh now published his writings in modern languages and not in Latin is in part consistent with his view, already presented in the Education Committee, of the supreme importance of living tongues. Of course he was also hoping for a wider circulation of his writings.
40 Berzelius to Agardh, 14 January 1828, LUB.

have been willing to check the *Textbook*.) Now, as the book had already been printed, nothing could be done to amend the Swedish version, but he was able to append a list of corrections to the German translation.[41] In his reply, Berzelius could not hide his irritation: The book contains many errors, and it is a pity that they cannot be rectified in the actual text of the German edition, as Agardh's circle of readers in Germany may suspect that the author has not paid due attention to awkward facts. Those who had studied the chemical aspects of plant physiology will be particularly critical of the book, as Agardh's basic theory of plant metamorphosis does not explain the fact that changes in form are normally accompanied by changes in material composition. Nevertheless, Berzelius asserted that the *Textbook* marks a new era in Swedish plant physiology, and he regretted only the *cave canem* that is "contained in the dedication to the most senseless of all naturalists," that is, Schelling. After all, Berzelius concludes, Agardh is not one of the speculative natural philosophers.[42]

Agardh did not seem to take this criticism very seriously, and in the German translation he made none of those alterations that Berzelius suggested were necessary. In the appended list of corrections, Agardh included only misprints and minor addenda of other kinds. He also particularly discussed an essentially favorable review of the Swedish version by the German botanist F. G. Eschweiler.[43]

When completing the second volume of the *Textbook*, Agardh took no more care to avoid printing a still faulty manuscript. He handed over the final proofs of the book to Berzelius only when it was ready for press. No accompanying letter has survived; possibly Berzelius received the proofs when Agardh visited Stockholm. This time, Berzelius delivered a furious, scathing criticism, contained in an extremely long letter, which earned a well-merited notoriety in Swedish history of science, although it was known to only to the two correspondents during their lifetimes.[44]

The letter includes many detailed comments and corrections, which are of two distinct kinds. The first and longest list of criticisms concerns matters of chemical and physiological fact, terminology, and pure scientific theory. The second one consists of attacks on Agardh's expressed or implicit philosophy of science and of nature, his tendency to speculate, his construction of analogies between plant and animal physiology, and his neglect of sound experience and established fact. In addition, Berzel-

41 Agardh to Berzelius, 4 November 1830, KVA.
42 Berzelius to Agardh, 18 November 1830, LUB.
43 Agardh, *Lehrbuch der Botanik*, Vol. I (Copenhagen, 1831), pp. 417f. Eschweiler's review appeared in *Annalen der Gewächskunde*, 1827, 4,1:48f.
44 Cf. Söderbaum, *Levnadsteckning*, Vol. III, pp. 67f; and Sten Lindroth in this volume, Chapter 1.

ius criticized Agardh's style, which, like everything else in this book, bears signs of haste and impatience.

Here are a few examples of the first kind of criticism. Agardh wrote in his book: "Ferric oxide always combines with acids," and Berzelius corrected: "Ferric oxide seldom combines with acids, and never with carbonic acid." Agardh rejected Dutrochet's terms endosmosis and exosmosis in favor of his own active fluid and passive fluid. With reference to such errors, Berzelius remarked angrily that he is surprised to see that even in matters close to Berzelius's own science Agardh does not hesitate, despite his superficial knowledge, to voice tendentious opinions.

Agardh's alleged errors of fact are only one reason for the irate tone of Berzelius's detailed criticism. Another reason becomes evident not from what Berzelius stated openly but from a closer examination of his various remarks. In the *Textbook* Agardh made many references to scientific authorities of different kinds and periods, of whom Berzelius himself is one of the most important. Some of these references are made in paragraphs that, according to Berzelius, contain grave errors, and some of the supposed authorities are people for whom Berzelius had little respect. He found himself not only misinterpreted and unjustly discredited by the errors in the text but also in bad company.

Agardh and Berzelius had in fact very different views of the nature of scientific progress and scientific authority. The difference becomes even clearer in Agardh's subsequent letter and in Berzelius's answer to it. In his reply – of which the tone, characterized by sorrow, wounded pride, but continuing friendship, has been well described, but the content ignored – Agardh defended his standpoints in the *Textbook* against even the most detailed of Berzelius's criticisms.[45] He tried to refute Berzelius's accusation of being poorly versed in scientific literature by quoting numerous authorities for his own views. To this defense Berzelius responded in a letter marked by a friendly tone but offering a repetition of his criticism: He may have underestimated Agardh's wide reading, but the faults and mistakes in the text remain faults and mistakes notwithstanding the authorities to whom they can be attributed.[46]

There is one point in this criticism that is especially deserving of note. Berzelius pointed out that some of the theories of various scientists quoted by Agardh are obsolete, "knowledge of nature having reached a higher level since then." To Berzelius, scientific progress is strictly cumulative; new facts are discovered, mistakes are corrected and science advances steadily. Agardh's approach was not unambivalent, but it bears at least

---

45 Agardh to Berzelius, 11 January 1832, KVA. On its tone, see Sten Lindroth in this volume, Chapter 1.
46 Berzelius to Agardh, 24 January 1832, LUB.

traces of a typically Romantic view of science as fundamentally the creation of a few brilliant men of genius, whose wisdom is eternal.

In this respect he was not at all as dogmatic as Hwasser, to whom the history of medicine largely consists of a few timeless giants, such as Hippocrates, Galen, and Sydenham, who are far more worthy subjects for young medical students' study than the vast majority of modern doctors with their narrow-minded empiricism.[47] Agardh was by no means as extreme, but if we pause to consider his attitude toward his authorities, we can see that his views were a long way from those of Berzelius. In reply to Berzelius's second critical letter, he briefly observed that, by neglecting those scientists to whom Agardh had referred, Berzelius seemed to be accusing "the most experienced scientists of Europe" of ignorance.[48] This remark presupposes a conviction that an outstanding scientist inevitably remains an authority, at any rate for a long time.

These divergent views of the nature of scientific progress bear witness to an even more profound difference in epistemology and the theory of science, a difference explored later. However, we must first look at the real effect of Berzelius's more detailed, factual criticism on the German translation of the second volume of the *Textbook*.

We have already seen that in his letter Agardh defended nearly everything that Berzelius had rejected as erroneous. Although his tone is very deferential at the beginning of the letter, a fact that has misled the commentators who have seen his answer as a surrender to Berzelius, he did not actually admit any real mistakes. It is therefore noteworthy that in the German translations he makes many, quite substantial, corrections in accordance with Berzelius's criticisms. The changes must have been made in a hurry, as the translation appeared in 1832. In not a few cases, Agardh defended his standpoint in his reply but nevertheless followed Berzelius's recommendations in the German edition. For example, when in the Swedish *Textbook* he declared that electricity is a cosmic element influencing vegetation,[49] Berzelius commented sarcastically that this flowery word cosmic is incomprehensible. To this Agardh demurred that electricity is just as cosmic as light. However, the German reader will not find the word cosmic in the translation.[50] Another case may be of more modern interest. Agardh had challenged Alexander von Humboldt's opinion that sulfur might be useful as a fertilizer by pointing to the destruction of the environment in the neighborhood of a sulfur factory. Berzelius objected that the threat to vegetation is not sulfur but sulfuric acid. Moreover, horticulture in the vicinity of the Swedish factory at Falun was outstanding. In the German

---

47 Cf. Liedman, *Israel Hwasser*, pp. 93f.
48 Agardh to Berzelius, undated [early in 1832], KVA.
49 Agardh, *Lärobok*, Vol. II, p. 58.
50 Agardh, *Lehrbuch der Botanik*, Vol. II, p. 32.

translation, Agardh said merely that the soundness of Humboldt's recommendation had yet to be proved.[51] In short, Agardh seemed to have been much more prepared to act on Berzelius's criticism in print than would appear from his letter.

The other type of criticism leveled by Berzelius concerns not facts but philosophy. Berzelius had uttered many harsh words about speculative tendencies in science, but he was never as bitingly eloquent as in this long letter to Agardh. He found that much of the *Textbook* was nothing but pure fiction, and a paragraph in which Agardh declared that the mechanistic standpoint is not sufficient to explain vital phenomena was dismissed as a "Paracelsiade," that is, speculation à la Paracelsus.

Specious analogy, so typical of Romantic natural philosophy and not rare in Agardh's book, made Berzelius furious. The stomach and the lungs are the poles in an animal in the same way as the leaves and the root in a plant, Agardh maintained. This metaphor is groundless, Berzelius objected, and in the case of the animal it is particularly "distasteful." When Agardh exclaimed, "Plants do not need to think, as the sun thinks for them; the sun is their soul," Berzelius retorted: "Truth, my friend, is the soul of science. She does not tolerate dreams, even those of genius." Agardh did not even attempt to answer criticism of this kind in his letter, nor did he make any corrections to the German version to accommodate Berzelius's objections.

Agardh's own attitude to Romantic natural philosophy was far from unambivalent, and the *Textbook* is the most Romantic and speculative of his studies. However, even in other writings and at other periods, when he was considerably more down to earth than here, his scientific and philosophical ideas were not at all as empiricist as those of Berzelius. Even if he did not use such wild analogies as the comparison of stomachs to roots, he was fond of theorizing and did so with conspicuous ease.

Evidently he was not prepared to discuss his views in this field with Berzelius, and except in the few letters concerning the *Textbook*, Berzelius also avoided the subject. To both, friendship and agreement on science policy and education were too important to be jeopardized by dissension in an area where their opinions were so strikingly different and, at the same time, so important to each of them.

After the long and intense exchange on the *Textbook* the Agardh-Berzelius correspondence calms down to brief letters, dealing mainly with routine questions concerning the Academy of Sciences or personal matters such as the future of Agardh's son. The tone is still friendly, if not as intimate as in the days of the committee. They saw each other in Stockholm, both as members of Parliament and in the meetings of the Scandi-

51 Ibid., p. 30.

navian Scientists, founded in 1839, where they both played leading roles. Here they found it possible to express their contradictory ideas of science freely and fluently, but without any suggestion of mutual criticism.

In his last letter to Berzelius, written in November 1846, Agardh enclosed his obituary of their common friend Esaias Tegnér and a new treatise on mathematics. Berzelius was by now an old man, but Agardh, the 61-year-old bishop, was entering his most productive and versatile period as an author.

Conclusion

Most historians of science who have dealt with the disputes between Berzelius and Hwasser or with the divergent opinions of Agardh and Berzelius on the proper approach to natural science have seen Berzelius as a representative of the rational, modern, and ultimately victorious tendency in history, and Agardh and, even more, Hwasser as in the camp of old-fashioned dreamers and visionaries. In the case of Hwasser, this verdict seems especially justified, as his campaign against Berzelius was not confined to the defense of the still backward university against the progressive young institute. For a period it also embraced the problem of how to handle the cholera epidemic of the 1830s. Here Hwasser promoted the miasma theory, that is, that the disease was spread by bad vapors, caused by earthquakes or volcanic outbursts, which was not only an outdated theory but also one that caused a lot of harm by implying that any hygienic measures taken by the authorities were useless. Berzelius opposed Hwasser here, too, proposing a contagion theory. As always, he refrained from mentioning his opponent by name.[52] In other words, Berzelius was one of the happy few who really pushed forward the frontiers of knowledge, whereas Hwasser tried to maintain obsolete opinions and dangerous prejudices. However, Hwasser did have some ideas that promised to be more fruitful and that which are still challenging. Confronted with the problems of modern, highly specialized, and compartmentalized higher education, we may find some inspiration in his dreams of a comprehensive university aiming at the true *Bildung* of all its students. The same applies to his vision of a teacher-student relationship characterized by humanity and affection!

As for Agardh, his ideas on education are now not only taken seriously but are even fashionable. Once again, the idea of a university that has both a deep social commitment and an autonomy that is inseparable from responsibility is being put forward, although never as eloquently as by Agardh. At the same time, the ethical standards formulated by Berzelius,

52 Liedman, *Israel Hwasser*, pp. 147f.

who wished to see the scientist pursue knowledge for its own sake without regard to its practical consequences, seem much more hazardous in a time of devastating pollution and an ever-present capacity for nuclear war.

Of course, Berzelius could not possibly have foreseen a future in which modern science would contribute not only to prosperity, relief, and enlightenment but also to destruction and anguish. However, it is no longer appropriate to see him as just one brilliant link in a wonderful chain of progress stretching from Galileo to the Nobel laureates of today. George Sarton's idea of a hagiographical history of science is no longer acceptable. Neither the history of religion nor the history of science has its real saints or its real sinners – only people who are more or less clever, ambitious, unprejudiced, far-sighted, versatile, charismatic, or stubborn. We must not overestimate the modernity of Agardh when he said that sulfur is harmful to the environment nor attribute any cynicism to Berzelius when he denied it. It is all too easy to recognize our own experience of education in a high-tech society when we follow Hwasser's polemic against a particularized vocational training. In fact, he was lamenting an Arcadian world that never existed.

We must also take care not to see Berzelius simply as the man of a new epoch and his more Romantic opponents as members of the ancien régime. History is much more complicated than that. In his day, Berzelius represented the era of the Enlightenment; Agardh and Hwasser, who were a good deal younger, were more modern in their closeness to Romantic modes of thought, which were a reaction against the Enlightenment.

History never really lends itself to immediate identification of ourselves and our own thoughts and feelings. The really fruitful historical experience is the experience of something that is very alien to us and yet fundamentally close to our own human existence.

3

Berzelius and the Atomic Theory: The Intellectual Background

GUNNAR ERIKSSON

Berzelius's relationship to the atomic theory has usually been analyzed in connection with the work of John Dalton. However, in this essay attention is focused on another aspect of Berzelius's relationship with the atomic theory, namely, the position taken by the celebrated chemist in the broader debate on the nature of matter that was conducted in his time not only by chemists but also by physicists and philosophers. The controversy was a fundamental stage in the lively discussion of different world views that was engaging men's minds in an era of rapidly shifting ideological currents. Here, too, the work and ideas of Berzelius have their undoubted place.[1]

Opposing Theories of the Structure of Matter

Classical Atomism

A glance at the state of the discussion on the eve of Dalton's and Berzelius's entry on the scene actually reveals a picture more complex perhaps than is usually realized. An important part of it naturally consisted of the ancient mechanistic atomism, which had its roots in the teachings of Leucippus and Democritus and which was handed down through the Epicurean school of philosophy to the Renaissance – with its predilection

Translated by Bernard Vowles from a revised version of Gunnar Eriksson, "Berzelius och atomteorin: Den idéhistoriska bakgrunden," *Lychnos* (Annual of the Swedish History of Science Society), *1965/1966*:1–34.

1 Berzelius's position in the philosophical controversy of his day has been depicted by Sten Lindroth in this volume (see Chapter 1). In the present essay, a shortened and revised version of one first published in *Lychnos*, 1965/66:1–34 (followed by an English summary on pp. 35–37), the background to some aspects of his concept of nature are presented in somewhat greater depth. Some secondary literature that appeared since the original publication of this essay is cited here, but it has unfortunately not been possible to give detailed consideration to all of the relevant material.

for all classical theories – and then to modern physics, to Gassendi, Boyle, and Newton, with whose authority it became a more or less fundamental tenet of the physics of the eighteenth century. This widespread atomism was often relatively vague in form.[2] We meet it frequently in textbooks from the late eighteenth and early nineteenth centuries.

Typical formulations appear in a well-known Danish compendium, Adam Wilhelm Hauch's *Inledning til naturkunnigheten* [Introduction to science], which appeared in Sweden in a translation by Carl Gustaf Sjösten just as the new century opened. The problem of the nature and structure of matter was raised, as was normal in the textbooks of the period, at the beginning of the book, where the "general properties of bodies" are dealt with.[3] According to Hauch these properties are spatial extension, impenetrability, porosity (with regard to the structure of macroscopic aggregates of matter), divisibility, cohesion, mobility, and weight. Like the old atomists, Hauch found that the physical world is dependent on two constituents. Besides matter itself, his description also postulates the existence of empty space, essential to explain the variations in density among different bodies. Although this basic conception of matter seems almost inevitably to lead to an atomistic viewpoint, Hauch introduced the concept of the atom rather hesitantly. He ascribed a fundamental role to divisibility as a characteristic of matter (a theme pursued here repeatedly), and this led him to pose the usual question whether this divisibility can continue ad infinitum. The true atomists denied that it can, as in their opinion the atoms have a quite definite corporeal extension, while at the same time being eternal and indestructible. Like many contemporary physicists, Hauch shied away from categorical pronouncements. He believed that whether divisibility reaches a particular limit or continues to infinity is a metaphysical rather than a physical question. No reason can be given for the existence of a particular limit, although "on account of the weakness of our senses and the imperfections of our tools" we hu-

2 Accounts of the history of the atomic theory that cover this period include L. L. Whyte, *Essay on Atomism* (New York, 1960) (brief but thoughtful); and A. G. M. van Melsen, *Atom gestern und heute*, Orbis academicus, Vol. II:10 (Munich, 1957). See also the article "Atom," in J. Mayerhöfer, *Lexikon der Geschichte der Naturwissenschaften*, Vol. III (Vienna, 1962). Interesting aspects of the status of the atomic theory in the period here treated may be found in Marie Boas, "Structure of Matter and Chemical Theory in the Seventeenth and Eighteenth Centuries," in *Critical Problems in the History of Science*, ed. Marshall Claggett (Madison, 1959), pp. 499–514; Thomas S. Kuhn, "Robert Boyle and Structural Chemistry in the Seventeenth Century," *Isis*, 1952, 47:12–36; and H. Kubbinga, "Le développement du concept de 'molécule' dans les sciences de la nature jusqu'à la fin du XVIIIe siècle," Thèse du doctorat de troisième cycle, École des Hautes Études en Sciences Sociales (Paris, 1982).

3 A. W. Hauch, *Inledning til Naturkunnigheten*, tr. from the second Danish edition by C. G. Sjösten, Vol. I (Stockholm, 1800), pp. 10–34.

mans must at some point stop dividing the bodies. Nevertheless, Hauch
was prepared to introduce the concept of the atom, in its then accepted
sense, into his account. Atoms, he stated, actually exist only in our imagination, but they are our designation for the "very smallest part of bodies
... with which we combine the concept of Impermeability, Extension
and Shape, without Porosity and Divisibility."[4]

Hauch's treatment of the atomic theory offers an illuminating example
of the often abstract character of the subject before the arrival of Dalton
– removed from the practical hypotheses used every day by physicists
and chemists as guidelines for their experiments. When he discussed the
difference between chemical and mechanical division and in so doing
introduced the concept of "element," he found no reason to commit himself on the question of atoms, nor did he bring it up when he came to the
question of chemical affinity. It was in just such contexts that the atomic
hypothesis was soon to prove so fruitful, but this was hardly suspected
yet by Hauch or by the majority of contemporary scientists.

One important reason for the doubts of Hauch and many others was,
of course, the lack of experimental support for the atomic theory. However, another reason is quite definitely to be sought in the very character
of Newtonian mechanics, which dominated the eighteenth-century view
of nature and its workings. It cannot be denied that Newton's system is
often ambiguous with regard to certain basic questions, which could be
overlooked in the more restricted research that makes up the scientist's
everyday activity but which arise in every detailed analysis of the assumptions on which the then accepted world view rested.[5] Such ambi-
guity attaches to the fundamental concept of force. Whereas a so committed corpuscular theoretician as Descartes considered that he had
produced an interpretation of nature that actually rested on a single
foundation, that everything in the world consisted of matter in motion,
Newton and his successors were compelled to introduce alongside matter
itself the theory of forces, on which matter was dependent.

Another problematic area among the prevailing concepts of science
was encountered in the imponderables, which many physicists and chemists thought they ought to take into account. These included magnetism,
electricity, heat, and light. These imponderables might to some extent
have been regarded by certain theorists as forces rather than matter, but
this did not make the theoretical foundation either firmer or more homogeneous. To the extent that they were seen as substances, they had
surely to differ in some essential way from ordinary ponderable sub-

4 Ibid., p. 17.
5 An extraordinary analysis of many of the more or less unconsciously accepted fundamental assumptions of modern science may be found in E. A. Burtt, *The Metaphysical
  Foundations of Modern Physical Science*, rev. ed. (London, 1932, with later reprints).

stances, and this issue might also have a direct bearing on the question whether all matter could be construed as built up of atoms. In Dalton's scheme of things the atoms of the different ordinary elements were surrounded by a sphere of heat (in the form of infinitely smaller atoms), and the same idea had already occurred to Robert Boyle. This image could scarcely help to make the atomic theory more acceptable as a theoretical basis for a unified concept of matter.

The Dynamist Conception

In these circumstances it is not surprising that many theorists proposed a totally different basis for understanding the structure of the physical world. Instead of regarding solid matter made up of some firm substance as the actual original constituent, they made the concept of force the ultimate foundation. If matter was ultimately not physical in the conventional sense but dependent on centers of force, it became easier to fit the imponderable substances into an integrated system, and gravitation and other forms of attraction merged with the idea of matter into a single unity. The concept of force might appear applicable to all physical phenomena, but matter on the other hand only to some of them; the remainder were left unexplained and inexplicable. Gottfried Wilhelm Leibniz developed such a concept of force in his system of monads, and the physicist Roger Joseph Boscovich presented a very advanced theory that developed the idea that the basic constituents of matter were spatially situated point centers of force. However, in the discussion of the nature of matter in which Berzelius took part, the relevant philosophical theory of forces was a later one, formulated by Immanuel Kant and generally referred to in the literature of the period as "dynamism."

In the principal surveys of the history of the atomic theory the importance of Kant in this context has scarcely been noticed, and certainly not given any prominence.[6] The picture of Kant normally presented in the history of ideas has been so much an image of the great critic who urged

6 Whyte, *Essay on Atomism*, p. 56, mentions Kant in passing. Mary B. Hesse, *Forces and Fields: The Concept of Action at a Distance in the History of Physics* (London, 1961), pp. 170ff, treats in detail other aspects of Kant's physical theories than those taken up here, but does not indicate the influence they exerted on many natural scientists. In his *Michael Faraday* (London, 1965), L. Pearce Williams points out how English scientists became acquainted with Kant's dynamism through the poet Samuel Taylor Coleridge; more recently, Coleridge's influence has received extended treatment by Trevor H. Levere in his *Poetry Realized in Nature: Samuel Taylor Coleridge and Early Nineteenth-Century Science* (Cambridge: Cambridge University Press, 1981). Cf. also H. A. M. Snelders, "Atomismus und Dynamismus im Zeitalter der deutschen romantischen Naturphilosophie," in *Romantik in Deutschland: Ein interdisziplinäres Symposion*, ed. Richard Brinkmann (Stuttgart, 1978), pp. 187–201.

moderation on philosophers and reminded us humans of the limitations of our thinking that the very speculative aspects also present in his work have, perhaps, been overlooked. The doctrine of the mature Kant was in reality not only a reminder to science that it cannot transcend the limited sphere of categorized experience allowed to us by our own cognitive capacity; but it also taught that philosophy can lend empirical science a helping hand with regard to the basic conception of nature. In his youth Kant had devised his well-known theory of the development of the planetary system. During his great philosophical period he formed his vision of the ultimate structure of matter, which had far-reaching effects on the immediate development of the intellectual life of Germany and which in Berzelius's day was a topic of heated debate. It signified a radical departure from atomism.

The work, well known by name but seldom read, in which Kant put forward his criticism of the materialist atomic theory and his diverging view of the nature of matter is *Metaphysische Anfangsgründe der Naturwissenschaft* [Metaphysical foundations of natural science], which was first published in 1786 and reprinted at least four times during his lifetime.[7] Kant holds that philosophy can assist science with certain fundamental definitions, a sort of fixed starting points that are both inescapable and yet impossible to reach by the route of experience. They concern the most general concepts of physics, such as space, matter, and force. A number of these concepts are now defined by Kant on the basis of his well-known categories, grouped into the four forms of quantity, quality, relation, and modality. The main object of his interest was the concept of matter, as the ultimate concern of science is matter. In terms of the four forms of category, Kant gave matter four corresponding definitions, each of them providing the basis of its own science: phoronomy, which is based on the quantitative category; dynamics, corresponding to quality; mechanics, corresponding to relation; and phenomenology, which deals with the scientific examinations of matter from the aspect of modality. In phoronomy, matter is equated with what is mobile in space, mechanics defines it as "the mobile as far as this has moving force," and phenomenology defines it as "the mobile as far as this can be the subject of experience."[8]

In our context these three definitions mean little, whereas the dynamic determination of the concept of matter is the very starting point of Kant's

---

7 See Otto Buek's introduction in *Immanuel Kants Kleinere Schriften zur Naturphilosophie*, Pt. 1, 2nd ed. (Leipzig, 1909), pp. xxxviiif. Hereafter references to this work by Kant refer to the unaltered second edition (Riga, 1787), the oldest available at Uppsala University Library.

8 Kant, *Metaphysische Anfangsgründe der Naturwissenschaft*, 2nd ed. (Riga, 1787), pp. 1, 106, 138.

discussion of atomism.[9] The dynamic definition of matter is, in all its seeming triviality: Matter is the mobile as far as it fills a space. A corollary immediately complicates this description, for here Kant added that matter occupies a space not through its mere existence but through a special moving force. The concept of force has thus been introduced as a necessary constituent of the concept of matter, and this step has extremely significant consequences. Kant also took pains to prove his corollary. Penetration of a space, he argued, is motion. Resistance to motion implies that the motion is reduced or changed into immobility. But a motion cannot be reduced or nullified other than by a motion in the opposite direction in the thing that is moving – this is one of the axioms of phoronomy. The resistance that a body offers to another body attempting to occupy its space is therefore the *cause of* such a motion in the opposite direction in the encroaching body. But another name for "cause of motion," Kant held, is just "moving force." Thus matter occupies a space not through its mere existence but through its moving force. With this thesis Kant had actually reduced the role of the "corporeal" in the physical world to nothing and instead placed all the emphasis on force itself. This shift becomes even more evident when Kant and his followers elaborated these basic physical principles.

The force whereby matter fills a definite space is repulsive. Repulsion prevents other bodies from taking possession of this space. The same force also makes a particular piece of matter impenetrable, a property that was more usually claimed by the atomists with regard to their smallest particles. Kant saw impenetrability differently. That a body should be penetrated by another body, he held, would mean that the spatial extension of the latter would be entirely abolished, which cannot happen. A certain force can admittedly compress matter into a smaller space than it at first occupied. A force greater than this can compress the same matter further, but for the compression to be carried so far that the matter in question did not occupy any space at all would require a force that is infinite, and no such force exists. Kant postulated that the repulsive force acts the more strongly as the space that is occupied by the matter within which it is acting diminishes. Here the difference becomes apparent between Kant's concept of matter and a form of atomism that was, at any rate, not unusual. According to the latter, atoms were perfectly hard and impenetrable. Kant's space-filling matter is indeed impenetrable, but not in an absolute sense, nor is its hardness absolute. In each individual case, in Kant's view, one can assume that a particular quantity of matter could be compressed into a space smaller than that which it at present occupies. Kant, therefore, called this property of matter relative or dynamic impe-

9 Ibid., pp. 31ff.

netrability, as opposed to the absolute or mathematical impenetrability assumed by the atomists.

One of the fundamentals of atomism was obviously the assertion that matter was not infinitely divisible. Atoms could not be split. It acknowledged in general terms that a geometrical figure could be divided into as many parts as one liked, but it drew a sharp distinction between this mathematical and abstract divisibility and a concrete physical one. Kant, on the other hand, claimed that matter is infinitely divisible – in his view there are no true atoms – for each part of a space filled with matter contains, as we have seen, a repulsive force, acting in all directions. Now the laws of geometry state that space is infinitely divisible, and so, the space that is filled with repulsive force is therefore matter. Thus, to Kant, the physical divisibility of matter is a consequence of the mathematical divisibility of space, a view wholly at odds with atomism. Matter makes itself felt as a force, and according to contemporary thinking in physics, force does not act at a point but continuously along straight lines.

However, matter does not, in Kant's opinion, manifest itself only through the repulsive force but also through its opposite, attraction.[10] These two forces were seen as the only conceivable ones, because forces can work only along straight lines and therefore can only attract or repel. Kant found it necessary to assume the attractive force, because the repulsive force, if it acted alone, would cause matter to expand perpetually and ad infinitum. In this way all space would actually become void and no matter would exist. One, therefore, has to assume that in every body there is a contractive force acting against the repulsion and setting a definite limit to the expansion of that body. Similarly, it may be said that matter cannot depend only on an attractive force. For attraction alone would cause matter to contract more and more until its volume became that of a mathematical point and it thus ceased to exist. One must always assume a repulsive force that sets a limit to the contraction. Only in combination can the forces explain the property of matter of filling a definite space.

From Kant's point of view, therefore, the dualism is invalidated that in Newtonian mechanics follows from accepting both bodies and forces as independent fundamental concepts. Whereas Descartes had concerned himself only with the corporeal, Kant arrived at his monism by considering only the forces, the dynamic element. The physical "solid substrate" that common sense wishes to associate with matter plays no part whatever in Kant's explanation of any physical relationship, at least not in dynamics. In his system it can simply be disregarded.

10 Ibid., pp. 52ff.

As has been mentioned, the prevailing atomic theory assumed not only atoms of absolute hardness and density but also a void in which these atoms moved, combined, and dispersed again. Kant believed that a void cannot be demonstrated, and he found that the hypothesis of the void gives a poor interpretation of the phenomena, one that ought not to be accepted unless all other attempts at explanation fail.[11] Admittedly, the theory that space is filled with forces does not entirely exclude the possibility of a void, for matter possesses attractive force in a space even when it does not entirely fill it. For a space to be entirely filled, according to his theory, there must be a repulsive force acting against the attraction. However, the possibility of a vacuum does not make it a reality. In Kant's opinion all known physical phenomena may be explained by assuming the combined action of the two forces in different ratios. He tried hard to find a secure anchorage in his dynamic model for some of the most obvious properties of observable physical bodies, well aware that the atomists had succeeded very well in setting the visible and measurable phenomena in the context of the invisible world, inaccessible to observation, of small, hard particles in the void.

However, Kant was extremely cautious when he approached the more specific physical phenomena. The task of the philosopher is not to explore them, but only to give the essential foundations for such an exploration. Nevertheless he implied that chemical phenomena, the combination of different substances with each other to form compound bodies, may very well be explicable in dynamic terms. Whereas the atomists – in a manner still undefined – assumed that such a combination implied the mutual juxtaposition of the particles of the constituent substances in fixed combinations, Kant imagined that chemical combination, at least when it is complete, involves actual intussusception (his term): The particles penetrate one another and form a new whole, constituted by the new proportion between the forces entailed by the combination of the forces of the original substances.[12] And his efforts to clarify on dynamist premises why different substances have different properties follow the same line. The differences between substances may be assumed to depend on the different strength of the repulsive force in different forms of matter, whereas the strength of the attractive force is everywhere the same. The same quantity of matter in the form of different substances will therefore occupy differing volumes of space, and conversely the same volume of different substances will differ in density. The ether, which is conceived as filling all space and thus occupies the place where the atomists had visualized a great void, is the concrete example on which Kant dwelled. It must be a substance whose repulsive force, relative to its attractive

11 Ibid., pp. 154ff.   12 Ibid., pp. 95ff.

force, is immeasurably greater than that of any other matter known to us.[13]

The Dissemination of Dynamism in Germany and Scandinavia

The basic dynamic philosophy thus formulated by Kant was assimilated by German Romantic nature philosophy and there radicalized and used as one of the main arguments for the visionary basic thesis of Romanticism: that everything was living spirit, that the antithesis between inanimate, apparently mechanically functioning matter and the domain of the soul was only illusory. Matter as explained by forces fitted into this picture, whereas the hard particles of Boyle and Newton did not. The concept of force is much more easily associated with the spiritual sphere than is the concept of body; indeed force and soul have often been synonymous. "For *force* alone is the *insensible* in objects," said Friedrich Schelling, "and the spirit can only be compared to what is analogous to *itself.*"[14]

Schelling's first formulation of dynamism came in *Ideen zu einer Philosophie der Natur* [Ideas toward a philosophy of nature] (1797). He naturally differed from Kant on certain points, and with him there is no sign of the caution that, when all is said and done, is always present in *Metaphysische Anfangsgründe*. More emphatically than Kant, Schelling stressed here that matter is nothing more than the forces working within it. To distinguish between matter and force is possible only in abstract thought, but can never take place in reality. He also saw the opposed forces as in a certain way identical; they have the basis of their existence in each other and comprise each other. Above all Schelling naturally differed from Kant in that he incorporated dynamism into his own metaphysics and ultimately derived it from his particular epistemological principle, intellectual intuition. However, in regard to those parts of the theory that might be of direct significance to a physicist or chemist, there is no great difference between Kant's and Schelling's earliest conceptions of it. Schelling later extended the theory, emphasized the points at which he differed from Kant, and introduced the law of gravitation as a third necessary and uniting fundamental force; but not even with these changes did his dynamism lose its similarity with the original model in regard to the crucial point that matter was seen as nothing other than a state of tension between forces.

Around 1800, then, there were two versions of dynamism, both of them topical and keenly debated. For while the Romantics argued with

---

13 Ibid., pp. 102ff.
14 *Ideen zu einer Philosophie der Natur*, in *Sämmtliche Werke*, Vol. I:2 (Stuttgart and Augsburg, 1857), p. 219.

enthusiasm for their interpretation of Schelling's physics, there were scientists who were more interested in the original doctrines of *Metaphysische Anfangsgründe* and whom there is hardly reason to call Romantics. This is an important point. The many scientists who came to scorn Romantic nature philosophy for its turgid speculation were not obliged, therefore, to include dynamism in their negative judgment. Although dynamism found its most devoted advocates among the Schellingians and although it largely served their view of nature, it also possessed its own influence outside the discussion of the Romantic alternative in the debate on the nature of matter.

It has sometimes been stated that *Metaphysische Anfangsgründe* attracted very little notice by comparison with Kant's other writings from his main philosophical period.[15] Nor, perhaps, was it so often quoted, at least openly, in philosophical debate, even though Schelling based important arguments on it. But among scientists it evidently did not go unnoticed. The detailed textbook of the Halle professor, F. A. C. Gren, in particular, must have served to circulate its ideas more widely among physicists. Entitled *Grundriss der Naturlehre* [Outline of natural science], it was one of the most respected and widely read physics textbooks in the late eighteenth and early nineteenth centuries. It has a special chapter in its third (1797) and later editions devoted to what Gren called "metaphysische Naturlehre" and contained nothing less than the main theses of Kant's work and a dynamistic criticism of atomism (with which he had previously been very much in sympathy).[16] And within a few years Kant's work appeared again in a new edition, which shows that there was at least a readership for it. The situation in Sweden was not different from that in Germany and Denmark. When the nature of matter was discussed at all in Sweden, dynamism offered an alternative to the vague atomism encountered in Hauch's textbook. The debate in Sweden took place during Berzelius's student years and his first period as an experimental and theoretical chemist. It can hardly have failed to make an impression on him.

The importance of Denmark to the intellectual life of Sweden at this period has perhaps never been sufficiently emphasized. An early and, in Scandinavia, influential supporter of Kant's dynamism, who at least when he first appeared took a line very independent of Romantic nature philosophy, was the Danish physicist Hans Christian Oersted. His early work, *Grundtraekkene af Naturmetaphysiken* [Fundamentals of natural meta-

---

15 See Buek's introduction in *Kleinere Schriften*, pp. xxviiiff.
16 F. A. C. Gren, *Grundriss der Naturlehre*, 3rd ed. (Halle, 1797), pp. 31f. Gren also touched on the dynamist theory in his frequently quoted *Grundriss der Chemie*, Vol. I, 2nd ed. (Halle, 1800), §16ff, but without discussing the intrinsic nature of matter in detail.

physics], which was published in 1799 and was actually a product of his
critical study of the book by Hauch, adhered closely to the terminology
and general opinions of *Metaphysische Anfangsgründe,* but devised its
own arrangement of the philosophical material and claimed to have succeeded in a more a priori manner than Kant in deriving the fundamental
forces from the ultimate principles of philosophy. This is not the place
for a detailed study of Oersted's natural philosophy, but it may be noted
that throughout his life he appears to have retained some essential part
of his first profession of dynamism, which in itself must have accorded
an important role to dynamism in the debate on matter both in Denmark
and in the rest of Scandinavia. A decade after *Grundtraekkene* we find
Oersted in a review of the history of natural science (physics and chemistry) joining Schelling and Henrik Steffens in once again praising Kant,
whose perspicacity has freed natural science from the atomist system.[17]
And in 1829, as we can see from his interesting correspondence on atomism and dynamics with his friend Christian Samuel Weiss, he was still
in agreement with this devoted dynamist that bodies occupy their space
not by their mere existence but by their activity, their forces.[18]

However, with time, Oersted's views developed into a striking compromise with the atomic theory. The opinions and findings of Dalton,
Berzelius, and Joseph Louis Gay-Lussac made a very strong impression
on him, and we find that, like them, he was inclined to assume that matter is not continuous but consists of extremely small parts, and that
chemical combination does not involve complete intussusception, the
particles there being discrete. But he did not thereby become an atomist.
When, in the early 1820s, Oersted evaluated Dalton's theory of proportions, he stressed that it was compatible with a basic dynamist position.
The "base part" – he preferred this term to the materialistically loaded
"atom" – of matter could be understood dynamistically if it were assumed that for every "force relationship," that is, for each specific substance, there were a minimum magnitude, below which this force relationship could not be expressed. When he poses the question where the
simple numerical ratios of the chemical proportions come from, he answers it in a dynamistic spirit: All the elements are created from a common matter that, in accordance with a definite law, has "contracted to
substances of individual degrees of density and directions of force."[19] As
early as 1812, in the important *Ansicht der chemischen Naturgesetze* [A

---

17 *Förste Inledning til den almindelige Naturlaere* (Copenhagen, 1811), p. 38.
18 *Correspondance de H. C. Örsted avec divers savants,* ed. M. C. Harding, Vol. I (Co-
    penhagen, 1920), pp. 280–340.
19 "Udsigt over Chemiens Fremskridt siden det Attende Aarhundredes Begyndelse," from
    Tidskrift for Naturvidenskaberne, 1822, *1,* quoted here from Oersted, *Naturvidenska-
    belige Skrifter,* Vol. III, pp. 313ff.

view of the laws of chemistry], he favored this path of compromise and argued for a "dynamic atomism."[20]

Two other Danish science textbooks that were translated into Swedish are relevant to our subject. Peter Christian Abildgaard and Erik Wiberg's *Inledning till Allmän Naturkunnighet* [Introduction to general science] was translated and published in 1802 by Olof Lindén. The term atom is not found here, any more than are the terms mechanism and dynamism. Nevertheless, it is easy to see that the book propagates Kant's concept of matter. Matter is said to be infinitely divisible and is defined as that which is mobile in space. It cannot be conceived, according to the authors, without a combination of attractive and repulsive forces. For if the contracting force acted alone, "then matter would draw together to infinity," and if on the other hand the repulsive force acted alone, matter would expand into infinity.[21]

Carl Gottlob Rafn had also embraced dynamism in his remarkable *Utkast till en Växt-Physiologie* [Outline of plant physiology], translated by Johan Henrik Olin in 1799. This work was reviewed in Abraham Gustaf Silverstolpe's *Journal för Svensk Litteratur,* and among the critical comments expressed was the accusation that Rafn failed to make a consistent distinction between the atomist and dynamist theories, and the reviewer asked, rather conservatively: "Might it not have been better to keep to the conventional construction, rather than so to forget oneself in applying the new one?"[22] The review is an example of the kind of discussion that may well have taken place in Berzelius's immediate surroundings. This journal was published in Uppsala, written by academics in the city, and edited by Berzelius's first curator.[23]

Nor could Swedish readers who stuck to Hauch's atomist textbook fail sooner or later to make the acquaintance of dynamism. In the second part, which was published in Swedish in 1807, Hauch had been obliged to discuss this theory following the criticism that had been leveled at his work, in particular by Oersted.[24] He had been faulted chiefly for following the atomic system instead of the dynamic. Hauch readily acknowledged that both interpretations are of interest and worthy of comprehensive study, but he stressed that neither can claim to be anything more than a hypothesis. However, the dynamic system does suffer, in his opinion, from a major weakness: It has never been presented with sufficient clarity, and in certain respects it is obviously inferior to atomism. He objected particularly to the excessively large number of assumptions that

20 Oersted, ibid., Vol. II, esp. pp. 155ff.
21 Abildgaard and Wiberg, *Inledning,* tr. Olof Lindén (1802), §§1, 19.
22 *Journal för Svensk Litteratur,* 1799, 3:741.
23 See Anton Blanck, "Berzelius som medicine studerande," *Lychnos,* 1948/1949:171ff.
24 Hauch, *Inledning,* Vol. II (Stockholm, 1807), pp. 642ff.

underlay it. In his view no repulsive force needs to be postulated, for everything arising from its action can equally well be interpreted as the effect of attraction from the opposite direction. He finds unreasonable Kant's hypothesis that an unrestrained attractive force would shrink matter to nothing, but his argument shows that he evidently does not understand that matter in the sense of "solid substrate" no longer has any meaning to the dynamists. His own counterargument assumes such an independent existence of matter alongside the forces. Offering many other objections he uncomprehendingly addresses a question to the dynamists to which he evidently thinks it impossible for them to answer yes: "Would then these forces determine the essence of matter?" At this question many dynamist sympathizers must have felt that they could regard these and other objections from Hauch with equanimity.

In textbooks that undoubtedly reached a Swedish readership, either in the original or in translation, there was, then, information on dynamism that shows that it was taken seriously and that there was discussion of its right to be considered an alternative to atomism. It had, moreover, a place in oral instruction. With admittedly varying degrees of enthusiasm it was presented to physics students at Uppsala University. Zacharias Nordmarck, a respected professor of the subject for four decades, from 1787 onward, gave what appears for many years to have been a relatively unchanging course in the foundations of physics, which began with a discussion of the essential construction and nature of matter. The heavily thumbed notes of several students have survived to bear witness to how he portrayed atomism and dynamism as the two conceivable alternatives in the question, with a positive emphasis sometimes on one system, sometimes on the other.

We may follow the notes of the future professor of astronomy Johan Bredman on Nordmarck's lectures in the spring terms of 1803 and 1804.[25] The lecturer quickly reached the problem of the divisibility of bodies and confirmed that no limit to it is definable.[26] That every spatial extension can be divided ad infinitum is, he said, well-known geometrical axiom: "As this infinite divisibility takes place in all directions, it must also belong to all bodies, because a body has extension." Nordmarck paused to criticize those theories that assume some kind of ultimate constituent that is not in turn divisible and cannot itself be assumed to have any extension, calling this the monad system, the concept of matter held by Leibniz and, perhaps, by Boscovich. He also rejected, citing his fellow Swede Samuel Klingenstierna, the French writer Bernard le Bovier de Fontenelle's assumption of *"quantités infiniment petites."* But having thus re-

25 UUB, A 219.   26 Ibid., first pagination, pp. 5ff.

jected the idea of the absolute indivisibility of matter up to a certain limit, he still allowed that we do not know "a true division ad infinitum."

Only here did he bring the traditional atomic theory itself into the discussion. It has long been assumed that there is a definite limit to the division of matter, because the forces that act on matter are incapable of splitting it up beyond a certain point. The bodies whose size corresponds to this point of division are what we call atoms. Among the adherents of atomistic or corpuscular physics Nordmarck includes Epicurus and also, curiously enough, Leibniz (did he mean Descartes or Newton?) and the eighteenth-century physicists Jean André De Luc and Georges Louis Le Sage, who had exploited the hypothesis "to the uttermost." The gist of atomism, according to Nordmarck, is that bodies differ from one another only in the position of their atoms and the spaces between them. He contrasted this atomism with Kant's dynamism, based on the assumption of the mutual affinities and the intensity of forces of attraction and repulsion.[27] "With this theory one avoids several unreasonable explanations given by atomism," the lecturer owned. He added an objection to dynamism that had already been advanced by Newton: If there are no indestructible, constantly similar particles that can build up generation after generation of living organisms of identical structure, then "Animals and plants ought never to remain in their genera and Species – but continuously degenerate into new families by chance changes in the relationship between the forces of Attraction and Repulsion."[28] But Nordmarck clearly did not regard this objection, so fascinating in historical perspective, as decisive. It answers itself, he added laconically, and unfortunately we cannot know whether he accepted that species and families can change with time (less likely), or whether he considered the ratios of the forces just as constant as the hard atoms, or whether he visualized some third solution.

The few natural scientists among the professors of Uppsala in the early eighteenth century appear to have been wary souls who were reluctant to comment on the ultimate questions of their disciplines. Undoubtedly Nordmarck was the boldest and most committed. In the toiling throng of docents and adjuncts there was perhaps a greater receptiveness to the new ideas on the nature of matter. Whereas the professor of chemistry Johan Afzelius seemed relatively seldom to have treated his audience to any deep probing of the ultimate significance of the chemical concept of matter,[29] his capable adjunct, Anders Gustaf Ekeberg, possessed a very

---

27 Ibid., pp. 11ff.   28 Cf. Newton's *Optics,* Query 31.

29 From one of the many extant manuscripts containing introductions to his own chemistry lectures (UUB, D 1476, "Lectures on Chemistry in General"), however, it is clear

different interest in that side of his science that sought to be something more than "a fabricating skill, a meticulously described technical tradition handed down from man to man."[30] Ekeberg was actually one of the most interesting Swedish followers of Kantian dynamism, arresting not least because he sought to relate the philosophical principles to a highly topical issue in modern antiphlogistic chemistry, the celebrated theory of the nature of affinity and chemical combination, put forward by Claude-Louis Berthollet in 1801 in *Recherches sur les loix de l'affinité* and in 1803 in *Principes de la statique chimique* and successfully opposed by Joseph Louis Proust.

In the autumn of 1806 Ekeberg lectured with rhetorical bravura on these subjects. An important link in his argument was that the dynamist theory of matter clashed with certain points in Berthollet, to the latter's disadvantage. Ekeberg aligned himself most closely with Kant. It is true that he implied that "among its more recent protagonists" critical philosophy had taken a new turn, but he thought that the scientific position at issue "remains on the whole unshaken, and that we are not venturing into anything new or unusual when we pin our theory to these propositions which have been received with both approval and admiration."[31] He cited the fundamental principles of dynamism in the customary manner and particularly emphasized its interpretation of the nature of chemical combination. This he called "chemical dissolution," explaining on Kantian lines that it is completely effected "when in the solution no separate part (in however small a space) can be found of any single one [of the starting materials], but however small a portion of the solution I visualize may contain both, in the same proportion as the whole. Thus both materials must together occupy the same space with continuity, and in consequence thereof have penetrated each other."

However, the largest part in Ekeberg's subsequent account is played by the concept of attractive and repulsive forces. He relates this to the theory of affinity (anticipated here by Gren), which invokes attractive forces of varying strength acting between substances. He presented a thorough and extremely penetrating analysis of Berthollet's concept of affinity, particularly as it had been interpreted by his two leading German commentators, Ernst Gottfried Fischer and K. J. B. Karsten. [32] The points

that for at least one period of his life Azelius was an atomist. Here he called the atoms "moleculae primitivae integrantes" and found that the atomic theory best explains a number of physical properties of bodies, such as density and softness.

30 UUB, D 1502: "1 Oct. 1806" (Lecture, autumn term, 1806), sheet 2. This manuscript contains Ekeberg's own draft lecture. For Ekeberg, see also Blanck, "Berzelius," pp. 172ff, and n. 6 in Sten Lindroth's Chapter 1 in this volume.
31 D 1502, sheet 2.
32 Claude-Louis Berthollet, *Über die Gesetze der Verwandtschaft in der Chemie,* Aus dem

he discussed in detail naturally include the well-known thesis that chemical combinations do not occur in fixed proportions of the constituent elements and also the theory that an element in solution immediately combines not only with the substance with which it has the greatest affinity but also with all the other substances present in the solution in quantitative proportions corresponding to the affinities, and the proposition that affinity depends more on the size of the chemical mass than on the intensity of the attractive force. The result of Ekeberg's analysis, which is based both on deductive reasoning and on selected concrete examples of the chemist's laboratory work, is that he rejected the claim of Berthollet's supporters that the new theory of affinity will overturn all earlier chemistry. While Ekeberg considered that Berthollet has enriched and deepened our understanding of the processes involved in the origin and dissolution of chemical compounds, he rejected several of the French chemist's central concepts, often quoting the dynamist theory and showing that it does not agree with the principles of Berthollet.[33]

Another convinced Kantian was the docent in astronomy, Jonas Brändström, who later became adjunct in mathematics and was honored with the title of professor in 1818, a year before his early death. Evidence of his dynamist loyalties is to be found in a dissertation written in 1809, dealing with the dynamist and atomist views of the impenetrability of physical bodies, and thus a central contribution to the debate on matter.[34] To judge from his definition of matter as that which is mobile in space, Brändström supported the Kantian version – how matter is constituted from the two counteracting forces and can therefore in theory be compressed to a mathematical point and thus is characterized by a merely relative impenetrability, in contrast to the absolute variety visualized by the atomists. Brändström depicted the latter as a pure *qualitas occulta*, which remains totally unexplained by the system. Brändström's chief objection to atomism was precisely that it takes for granted the fundamental concepts themselves, the very concepts that he considered most urgently in need of investigation. Motion thus becomes an unexplained premise beside and quite independent of the atomist concept of matter. In dynamistic physics, however, motion follows from the concept of mat-

---

französischen übersetzt . . . von Ernst Gottfried Fischer (Berlin, 1802); Karl Johann Bernard Karsten, "Ueber Berthollets chemische Affinitätslehre," [Scherer's] *Journal der Chemie*, 1803, 10:135–156.

33 Ekeberg also mentioned the dynamist view when introducing his private lectures in the spring terms of 1809 and 1811 (also in D 1502), in which he considered the distribution of the substances in a chemical compound, but he gave no clear indication here whether he himself still shared the dynamist view.

34 J. Brändström, *De Atomistica et dynamica impenetrabilitatis notione*, dissertation, Uppsala, 1809, defended by N. O. Alner.

ter: Force is the common denominator of both the material vehicle and its motion. What Brändström did find appealing in atomism is its clarity, and he emphasized that he found a purely atomist system better than the current mixture of mechanical and dynamic philosophies.

Brändström's work thus demonstrates an important point to which we had cause to refer earlier. One weakness of contemporary physics was that it worked with basic concepts that were poorly defined and therefore also, what is more important, poorly thought out. Matter, force, and motion were assumed to be self-evidently what a tradition of sound common sense seemed to say they were; the conventional physicist weighed and measured without really thinking about the object to which his weights and measures were being applied. The younger generation was inclined to see this vagueness as unsatisfactory and demanded clarity. This was what Kant wanted to provide, as also did Schelling and the Romantic philosophers, however paradoxical this may appear in a movement that has become notorious for its obscurity.

That this Romantic nature philosophy soon found its devotees in Uppsala is well known. An interesting example of how dynamism could be derived from it is given by the docent in theoretical physics Per Schönberg in his thesis *De conjunctione chemica* (1813).[35] We have now reached the time when Dalton's atomism was beginning to become known and to exert a great influence on chemistry, whereas previously the questions involved had been primarily the concern of physicists. Schönberg, like Ekeberg, referred to the debate started by Berthollet and looked in particular at the counterarguments of Joseph Louis Proust and his successors to support the view that chemical elements enter into their compounds in definite proportions. Schönberg also went a step farther and discussed the problem of multiple proportions. Like Brändström and many others, he began with the atomist and dynamist definitions of matter. He observed, in line with Brändström, that both methods of explanation use the concept of force to explain how bodies cohere, but naturally he found this concept of force applied very differently. The dynamists do not distinguish matter from forces and believe that it is a product of the conflict between definite forces, so that in their system these forces represent the principle of the existence and essence of matter. In the atomist system, on the other hand, they only represent the agent that affects the atoms and forces them together. Here, then, they are only the principle of motion, whereas in dynamism they are the principles of both motion and

35 Schönberg died tragically young while traveling abroad in 1822. He had by then become an adjunct in mathematics and natural philosophy. On Romantic science in Sweden, see my essay "Romanticism in Scandinavia," in *Romanticism in National Context*, eds. Roy Porter and Mikulas Teich (New York: Cambridge University Press, 1988), pp. 172–190.

existence: Through them arise both bodies and space (this is the Romantic's view but not Kant's). Schönberg implied that different epistemological theories underlie the two interpretations as they appeared in his time: To the atomists, bodies are external objects, quite distinct from our cognitive faculties, *intelligendi facultas;* to the dynamists no real objects exist other than to the being who is equipped with reason (*intelligenti*), and without this reason they are incomprehensible.

Only the dynamist interpretation appeared to Schönberg reconcilable with a considered philosophical viewpoint. He concentrated particularly on the concept of force: For reason, the existence of matter does not become comprehensible unless one supposes that it has its basis in forces. Essentially, therefore, a body, whether simple or composed of several substances, is the same as a fixed relationship between the forces that have produced it. (However, the workings of these forces cannot be understood a priori, he conceded, but must be established by experience.) Compound bodies are not composed of atoms that have conglomerated, but chemical combination has occurred by intussusception, which means that in contact with each other the forces of the reacting substances have entered into a new common state of combination. The so-called combined substance has therefore arisen as a result of the occurrence of a new relation of tensions. If complete combination has been possible, which Schönberg believed is only an ideal case, it is homogeneous down to its smallest parts and infinitely divisible into parts of the same kind.

A fleeting glance at the University of Lund shows that a similar discussion of the nature of matter also occurred there. It should be borne in mind that Lund maintained strong ties with Copenhagen, whose university was often visited by young Lund graduates for study under eminent professors or in search of rarer literature.[36] There was, therefore, ample opportunity for those from Lund interested in chemistry and physics to come into contact with Oersted, who was fascinated by the problems we are discussing.

The subject of matter forms the main theme investigated by the docent in philosophy, Petrus Dahl, in the dissertation *Doctrina atomorum* (1812), which amounts to a criticism of the atomic theory and a plea for dynamism as propounded by Kant and, more particularly, Schelling. Dahl ranked the contributions of the two philosophers in the phrase *"Kantius bene, optime cel. Schellingius."*[37] Dahl dealt at length, inter alia, with how dynamism can explain the density of bodies as a consequence of a force of repulsion acting with differing strength in different substances. Dahl was the nephew of the elderly, widely read professor of philos-

36 Cf. Gunnar Eriksson, *Elias Fries och den romantiska biologien* (Uppsala, 1962), p. 106.
37 Dahl, p. 7.

ophy Matthias Fremling, who lectured in the autumn of 1811 on Schelling's *Einleitung zu dem Entwurf eines Systems der Naturphilosophie* [Introduction to the outline of a system of nature philosophy], which he reviewed systematically, page by page.[38] In so doing, Fremling must have supplied a good deal of information on dynamism, which occupies a prominent position here as in all Schelling's works on nature philosophy. A third example of the dynamistic sympathies in Lund is provided by Jonas Brag, professor of physics and astronomy, who was strongly sympathetic to Oersted's general view of science. In his lectures in the spring of 1816 he talked, inter alia, about the dynamic system and the "foundations of speculative realism," and in 1818 he compared the two theories of matter and took up "the difficulties of the atomic system."[39]

The Schools Clash: Svanberg versus Arrhenius

A scientific skirmish between the old school and the Romantics that has attracted a good deal of attention was sparked by an article in *Svensk Literaturtidning,* the organ of the Romantics, in 1813 inspired by A. W. Schlegel and entitled "Öfver den moderna Naturkunskapen" [On modern science]. In his presidential address to the Royal Swedish Academy of Sciences the following year, Jöns Svanberg, newly appointed professor of mathematics at Uppsala and formerly the holder of science posts in Stockholm, attacked the article.[40]

The actual object of Svanberg's address was to indicate how a real nature philosophy, a perfect knowledge of nature, should be constructed, and how it could be achieved. Svanberg dreamed that physics and chemistry might one day be capable of expression in mathematical terms and that it would then become apparent that the many seemingly disparate phenomena described by these sciences could be understood as modifications of the same universally acting basic forces. In this connection he attached great weight to the experimental study of the attractive and repulsive forces that are active in bodies and that explain their elasticity and density, cause them to cohere, act between them in chemical reactions, and so on. This emphasis undeniably indicates that Svanberg could feel sympathy for essential parts of the dynamists' train of thought; but at the same time he expressed such a distaste for every facet of Romantic nature philosophy that nothing that it stands for seems to have been acceptable to him.

38 Eriksson, *Elias Fries,* pp. 112f.   39 Ibid. p. 125.
40 For this controversy, see G. Ljunggren, *Svenska vitterhetens häfder,* Vol. IV (Lund, 1890), pp. 502ff; Sten Lindroth in this volume, Chapter 1; and Eriksson, *Elias Fries,* pp. 85ff.

At the end of his address he briefly referred to his more concrete objections to the new opinions, and he also touched on the conflict between atomism and dynamism. His position is interesting not least because he did *not* profess allegiance to the former. Instead he declared that "to true *Physici*" the distinction is reasonably unimportant. Svanberg evidently thought that as far as mathematics is concerned, a scientist can use any system at all to produce the same quantitative description of the phenomena. The "wretched utterances" of the Romantics on the dynamic nature of matter depend, he believed, "on a concept of force which is as yet purely Metaphysical, but which will probably have to become a Mathematical one" to demonstrate any potentially significant difference between the two systems

It is not immediately easy to understand Svanberg's way of expressing himself, which is laconic and oblique, as is indeed his address as a whole. One has to assume that among other things he wished to say that the forces of which the dynamists speak must be shown to be measurable and therefore capable of being submitted to mathematical analysis before they can profitably be applied to the problem of the structure of matter. Just how profoundly attracted he seems to have been by at least certain aspects of dynamism is shown by another adverse comment he offered on the Romantic nature philosophers: The honor of having discovered the distinction between atomism and dynamism belongs not to the new school but to the German-French philosopher and scientist Johann Heinrich Lambert and "as far as that goes, Newton will always remain (discounting one or two casual expressions) the first, just as also he has been the deepest thinker of all Dynamic Nature Philosophers."[41]

Svanberg received a rejoinder well known in Swedish literary history from a keen supporter of the new school, Carl Magnus Arrhenius, an otherwise little-noticed Stockholm civil servant. In the pamphlet *Om den Falska Analytiska Constructionen i Mathematiken* [On the false analytical construction in mathematics] (1814), Arrhenius examined a number of Svanberg's arguments and rejected them indignantly. That Svanberg could see no difference between the atomist and dynamist approaches was to Arrhenius's mind absurd. To him the difference in principle was perfectly clear: Atomism was the system for the pure empiricists, for those who accepted the existence of matter and motion without worrying about how in turn their existence was to be explained. Like Schelling and Brändström, he asserted that he saw Le Sage's pure atomism as a more consistent and honest system than one with an admixture of dynamist concepts, which in his opinion was so common among physicists and of which Svanberg, too, had given an example. Dynamism, the theory of

41 Svanberg, *Öfver begrepet of Natur-Philosophie* (Stockholm, 1814), pp. 33f.

the nature philosophers, was, on the other hand, a way of considering matter and motion that led to an understanding of the very basis of their existence. If forces were taken as the origin, motion became a consequence of their activity (which of course consisted in attracting and repelling), rest became a special case of motions arising in this manner and acting on each other, and matter finally became a consequence of the relative state of rest of such forces.

Arrhenius was of course a dynamist of the Schellingian school, not directly of the Kantian. He therefore attacked Svanberg's argument about *two* forces constituting matter, because, like Schelling, he assumed that a third force must inevitably belong to matter, the force of gravity bridging the opposites, seen as something quite different from attraction.[42] He also indicated, with Schelling, that the polarized forces act, with differing potencies, as magnetism, electricity, and chemical processes. He joined his mentor in thinking that a full explanation of all the different qualities that we perceive in bodies is given only by considering all these various manifestations of the basic forces, whereas the simple relationship of attraction and repulsion of which Svanberg had spoken really covers only our concept of the density of matter.

Here Arrhenius finds cause to praise Berzelius, however surprising this may appear. He found in particular that the Swedish chemist had added depth to the concept of affinity derived from Lavoisier. Lavoisier contented himself with the explanation that chemical reactions were due to this affinity but never went on to clarify what this affinity actually was. Here, Arrhenius says, Berzelius has stepped in with his electrochemical theory. "With that," observed Arrhenius, "Berzelius has given affinity what Lavoisier never could, a true physical meaning, or, in other words: He has shown that a living Sympathy, not merely suspected but actually experimentally revealed, prevails everywhere in Nature; and that the relationship of the Sun and the Earth, Light and all the hitherto incomprehensible phenomena of Fire are results of it."[43]

Berzelius

Berzelius cannot have felt altogether at ease about this approbation. But where in fact did this great researcher himself stand in this dispute?

In the celebrated dissertation *Försök rörande de bestämda proportioner, hvari den oorganiska naturens beståndsdelar finnas förenade* [Experiments on the definite proportions in which the constitutents of inorganic nature are combined] (1810), Berzelius refers to Dalton's hypothesis, of which he as yet had only secondhand knowledge from an article in an

42 Arrhenius, pp. 39f.   43 Ibid., pp. 52ff.

English scientific journal.[44] Here he presented only the results of his experiments on chemical proportions – he did not link the regularity that he had discovered to a theory of atoms. But it appears fairly certain from his allusions that he was a relatively convinced atomist. He wished only to present facts and the results of his own analyses, but the theory, he continued, "which is implicit in the results of the experiments stares you in the face, and the ideas to which these can also give rise will surely be aroused in every attentive reader, without any action from me."[45]

In the second part of his *Lärbok i kemien* [Textbook of chemistry], which appeared in 1812, he explicitly and approvingly mentioned the atomic theory. He harkened back to his predecessor William Higgins and to John Dalton, whose work he had still not read when he wrote this section.[46] However, in the same year Dalton sent him a copy of his work, and the two chemists exchanged a few letters, although without being able to share each other's views on quite a number of concrete details. Dalton was unwilling, for example, to accept Gay-Lussac's studies of the simple volumetric relationships that are characteristic of the chemical compounds of gases as support for his atomic theory.[47]

A year later, however, Berzelius could no longer give unqualified approval to the atomic theory.[48] On the contrary, he found, in his "Essay on the Cause of Chemical Proportions and on Some Circumstances Relating to Them" (published in Thomson's *Annals of Philosophy*, 1813 and 1814), that the atomic theory had difficulties to which he himself could see no solution. He began his discussion of those difficulties by stating how he himself viewed Dalton's atomic theory, to which he would evidently have liked to subscribe if the difficulties did not arise. By "atoms" he said that he understands the smallest corpuscles into which matter can be divided. He stressed at once that he does not wish to discuss whether matter is infinitely divisible or not (we have seen that the question was certainly a topical one) but he assumed only than an atom is *mechanically* indivisible and that a part of an atom is inconceivable. He had a strikingly clear mental picture of the appearance of these atoms; spheri-

44 Berzelius's dissertation is included in *Afhandlingar i fysik, kemi och mineralogi*, 1810, 3:162–276. See also Söderbaum, *Levnadsteckning*, Vol. I, pp. 322ff; Anders Lundgren, *Berzelius och den kemiska atomteorin* (Uppsala, 1979), pp. 96–114; and Evan M. Melhado, *Jacob Berzelius: The Emergence of His Chemical System*, Lychnos-Bibliotek, Vol. 34 (Madison and Stockholm, 1981), pp. 169–170, 168n, 181n, 202n–203n, 209.

45 Berzelius, *Förök*, p. 166.

46 *Lärbok i kemien*, Vol. II (Stockholm, 1812), pp. 559f, 576.

47 Söderbaum, *Levnadsteckning*, p. 326.

48 Ibid., pp. 449f; for a full account, see Lundgren, *Berzelius och atomteorin*. See also Melhado, *Jacob Berzelius*, pp. 270–278; Anders Lundgren in this volume, Chapter 4; and Tore Frängsmyr, "Berzelius som filosof," in Frängsmyr, *Vetenskapsman som hjälte: Aspekter på vetenskapshistorien* (Stockholm, Norstedt, 1984), pp. 125–136.

cal in shape and all equal in size, irrespective of which element they make up. This latter assumption is necessary to understand that combinations of atoms will have a regular appearance. He also considered that such combinations – the fundamental constituents of chemical compounds – arise as atoms of different elements gather into small groups where they lie in direct contact with each other. In this direct contact they are held together by electrical attractive forces in accordance with his electrochemical theory. He called these combinations of atoms, which in fact correspond with what we call the molecules of chemical compounds, "compound atoms of the first order." These can in turn combine with each other to form "compound atoms of the second order." Berzelius's picture of these relationships supposes that in the compound atoms of the first order one of the elements is represented by a single atom and that all the other atoms in the group must be in direct contact with this one. As in addition all the atoms are spherical and of equal size it follows from the laws of geometry that such an atom cannot contain more than 12 atoms of the other combining substance. The compound atoms are considered to be as mechanically indestructible as the simple ones.[49]

The important property of mechanical indestructibility forms the main reason that Berzelius needed to imagine that one of the constituent substances in the compound atom is represented by just one single atom. It defies "sound logic" to imagine anything else, "for in such a case there is no obstacle, either mechanical or chemical, to prevent such an atom from being divided, by means purely mechanical, into 2 or more atoms of more simple composition."[50] Insisting on a single atom also served Berzelius as a safeguard against what he feared would be the destruction of the theory of chemical proportions.[51]

It is worth pointing out that this Dalton-inspired view of the nature of atoms and compound atoms still departs in several respects from Dalton's own. Dalton himself criticized Berzelius's "Cause of Proportions," maintaining that he does not think that the atoms in the combined groups

---

49 Berzelius, "Cause of Proportions," p. 446. The textbooks of the period distinguished between mechanical and chemical division. Chemical division takes place in chemical reactions; mechanical division as a result of crushing, grinding, tearing, and the like. Berzelius thus attributed to Dalton's atoms only mechanical indivisibility and could therefore claim that compound atoms are also indivisible.

50 Ibid., p. 447. However, on exactly what foundation he placed this assertion Berzelius did not say.

51 As Melhado points out in *Jacob Berzelius*, pp. 276–278. The saltation in composition between distinct compounds that characterized much of inorganic matter would give way to the transitions between barely distinguishable types that characterized natural history. To prevent arbitrary proportions from degenerating into transitions, Berzelius found it useful to limit speciation by insisting on the single atom.

actually touch one another,[52] as each of them is surrounded by a sphere of heat. Nor, according to Dalton, are they themselves necessarily round, for the roundness is a property that belongs largely to the envelope of heat. As a consequence of this view Dalton did not need to conceive of all kinds of atom as equal in size, nor could he see why combinations of atoms could not occur in which all the constituent substances are represented by more than one atom.

Working on the relatively stringent assumptions that he found necessary, the Swedish chemist thus encountered problems in the course of his experiments that led him now to consider the atomic theory dubious.[53] The analysis of certain chemical compounds indicated that some oxides contained one and a half atoms of oxygen, a situation that was patently absurd if one accepted the corpuscular theory. In organic nature, moreover, it was possible to find compounds that seemed inevitably to contain combinations of one atom with more than 12 others. These difficulties made him inclined to refrain, at least for the moment, from speaking of "atoms" and to use other, less binding expressions. Following Gay-Lussac's studies of the combination of gases he used "volume" for the quantitative unit in the theory of proportions.

However, the concrete form in which Berzelius imagined these volumes is difficult to understand from his sporadic references; this volume theory appears as obscure as his own atomic model was lucid. Typically enough Dalton said in his polemic that he preferred to suspend judgment on it until it had been further developed.[54] Berzelius proceeded from the fact discovered by Gay-Lussac that gases combine with each other such that either their volumes are equal in size or the volume of one gas is an integral multiple of that of the other. It is clear that his law could not be reconciled with the atomic theory until scientists had accepted Avogadro's view that many gases occur in molecules, but evidently to Berzelius it seemed at this point only to be a splendid confirmation of the doctrine of multiple proportions. However, the observation of these volumetric relationships applied to gases; were it to be extended to include all substances, then it would apply to them in gaseous form, a state of aggregation in which many of them defied any method of quantitative examination known at that time. How, then, did Berzelius imagine that these "volumes" were constituted? Evidently they do not need to be thought of as indivisible. "In the theory of volumes," Berzelius asserted, "we can figure to ourselves a demi-volume, while in the theory of atoms a *demi-*

52 J. Dalton, "Remarks on the Essay of Dr. Berzelius," *Annals of Philosophy*, 1814, 3:174ff.
53 Berzelius, "Cause of Proportions," pp. 447ff.
54 Dalton, "Remarks on the Essay of Dr. Berzelius," p. 180.

atom is an absurdity."[55] Volumes do not constitute "hard spheres," but are described vastly more vaguely.

It is tempting to draw a link with the dynamist conception of matter. Nothing seems to prevent us from applying the volumetric theory to matter as conceived by the dynamists. One of the dynamists' main objections to the atomic theory was that it assumed small bodies of absolute impenetrability. This objection seems inapplicable to the volumetric theory, which leaves the question of the essential structure of matter open, whereas atomism does not. It is true that Berzelius did not utter a word to suggest that he may have been influenced by or made allowances for dynamism. The arguments that he advanced against the atomic theory are of an entirely empirical nature and relate to his very decided view of how an atomic matter must be constituted. However, his silence does not have to be construed as total indifference to the views that made themselves so strongly felt in the scientific milieu in which he worked. We know that Berzelius had the deepest scorn for the Romantic nature philosophy, and he must have been determined to have avoided the slightest suspicion of having any truck with it. We must also bear in mind that his essay was published in an English journal and that he considered that the English generally had not taken much notice of the German controversy on matter, which would have rendered allusions to it meaningless to the majority of his readership.[56]

Berzelius's doubts about the atomic theory were temporary. A couple of years after the essay in the *Annals of Philosophy* he once again professed himself open to atomism. He also showed, more clearly than previously, that he was well aware that the atomic theory also occupied a place within philosophical debates. This awareness first appeared in a dissertation on phosphoric acid in Gilbert's *Annalen der Physik* (1816), and then, in the same spirit, in the third part of his own chemistry textbook, published in 1818.[57] He brought up the discussion of atoms with the express proviso, typical of Berzelius, that he was treating a question of theory, a theory that must take into account all known experimental data and be corrected as soon as anything contradicts it; but he nevertheless gave a quite clear and detailed picture of his vision of nature, going

---

55 Berzelius, "Cause of Proportions," pp. 450f. For Berzelius's use of the volume theory, see Lundgren, *Berzelius och atomteorin*, pp. 114–123, 131–135.

56 In his *Lärbok i kemien*, Vol. III (Stockholm, 1818), pp. 41ff., Berzelius compared more closely the volumetric theory as he now understood it with the atomic theory, without expressing himself with much greater clarity. Certain passages indicate that he supposed that the smallest conceivable volumes of the elements would be in gaseous form and therefore scarcely impenetrable.

57 See Söderbaum, *Levnadsteckning*, Vol. II, p. 32; and Lindroth and Lundgren in this volume, Chapters 1 and 4, respectively.

far beyond the limits that could be reached with the crucibles in the laboratory. It is essentially the same atomic theory that he earlier found difficult to accept; he has only given way on his many reservations regarding the concept of compound atoms, previously the main cause of his misgivings.

What Berzelius now found most probable and most consistent with our experience, as he wrote in his *Lärbok,* is that material bodies consist of small particles that may be called atoms, particles, molecules, chemical equivalents, or something similar. Like most others, Berzelius preferred "atom." He ascribed to these atoms the same definite characteristics as earlier: They must be indivisible, for they must always be identical in size and weight. Berzelius still did not speak of mathematical indivisibility, stating expressly that he considered only a physical or mechanical indivisibility. The implication of the indivisibility of atoms is, he said, "that when a body has been mechanically divided to a certain degree, the continuity of the resulting parts cannot be further interrupted by any mechanical force, i.e., it depends on a force which is stronger than all the forces by which mechanical division is produced."

These atoms are far too small to be measured; matter is divisible until each portion is so small that it eludes our senses. We cannot, therefore, pronounce with confidence on the form of atoms, but to Berzelius it seemed highly probable that the atoms of simple bodies (the elements) are spherical, "because this is the form preferably chosen by matter, when it is not subject to the effect of extraneous forces." The molecules of chemical compounds – what Berzelius still called "compound atoms" – cannot, on the other hand, be spherical, but must have some other definite form "according to the number of the simple atoms and the way in which they gather together." Atoms of different elements *may* conceivably differ in size, but Berzelius found it most likely that they are all of the same dimensions, for the compound atoms always show a geometrically regular form, which ought not to occur if the constituent simple atoms were of differing sizes. This is as far in this hypothetical direction that the cautious Berzelius dared go.[58]

Berzelius visualized atoms as small, hard lumps, held together in larger or smaller groups by the electrical force (or the electrical fluid) that constitutes chemical affinity. We recognize distinct elements of the classical atomic theory in this picture, just as we see details for which Berzelius alone is responsible. His atomism belongs to the Newtonian tradition – to describe the nature of matter he makes use of two distinct basic concepts: matter that apparently exists independently and is not reducible to any other form of matter and forces with which this matter is endowed.

---

58 *Lärbok i kemien,* Vol. III, pp. 19ff.

Perhaps a trace of the influence of dynamism can be discerned in this prevailing atomism. If so, it lies in the explicit pronouncement that the indivisibilty of atoms applies only in a mechanical respect. Whether something other than mechanical forces can split the atom is a question that is left open; it is not an intrinsic part of the atomist tradition. But generally Berzelius repudiated dynamism. He did so in a manner that shows that he had familiarized himself with its basic ideas.

In the dynamic system, said Berzelius, speculative philosophy presents matter, "not without some premonition of what is correct," as a product of the striving of two opposing forces. Without mentioning Kant or his disciples among the physicists by name, he cited the dynamist doctrine of the merely relative impenetrability of matter: "If one of these forces totally dominated the other, the matter of the whole universe could be gathered at one single mathematical point." A consequence of this conception, in Berzelius's view, is the dynamistic idea that bodies in chemical combination penetrate each other (what Kant calls intussusception) "and that the mutual neutralization of properties that results from combination in most cases consists in this penetration." It is this theory of penetration that Berzelius saw as the most empirically problematical element of dynamism. In his view it conflicts with the theory of chemical proportions, which it has not in any way anticipated. Berzelius did not say explicitly that dynamism is in conflict with it, but he asserted that if chemistry had been led along by dynamism, "particularly in the course it has taken over the last three quinquennia" – he was referring here to Romantic nature philosophy – then the chemists would have failed to notice these important phenomena.[59]

Berzelius was very careful to distinguish between what he regarded as established fact and what served only as guiding hypotheses. But this concern did not mean that he shied away from hypotheses. Actually he showed quite a strong liking for visionary insights into the still undiscovered; in that sense he, too, was a typical product of his age. Even his atomic theory, to which he seems to have adhered in more or less unaltered form for most of his life, clearly contains things that cannot be derived absolutely strictly from empirical data. Consider, for example, his idea of the round shape and identical size of the atoms. However, atomism does not exhaust his views on matter. The imponderable substances – light, heat, electricity (which was also presented as the force binding the atoms), and magnetism – remained to be fitted into his system. He worked on this task while his atomic theory was taking shape.[60] Whereas the Romantics found it relatively easy to include these phenom-

59 Ibid., pp. 18f.
60 In his textbook he devoted separate sections to these phenomena. See, for example, Vol. I, 1808, pp. 8–85; 2nd ed., 1817, pp. 7–139.

ena in their general theory of forces, to Berzelius the atomist they seemed to belong to a different order from the rest of nature. He did not refrain from guesses about them; on the contrary, his imagination flourished here as, perhaps, nowhere else.

Berzelius was most inclined to see the imponderables as material substances in the traditional manner, not as oscillatory motion or the like. But they had in that case a characteristic that distinguished them from all other bodies – they were not affected by the force of gravity. Berzelius was prepared to accept this fact as a part of his conception of matter. He saw no logical contradiction in the idea that certain substances were exempt from universal gravitation. In the second volume of his textbook (1812) he pointed out another of their distinctive characteristics: They did not occupy space in such a way that it could not be occupied by another body; in other words they possessed an almost dynamic quality.[61] (When he sought in the second edition of the first part of the textbook, 1817, to characterize magnetism, he clearly showed himself shifting between a corpuscular and a dynamic conception: "In every body there is a substance undetectable to our senses, like the electrical, which consists of two different and opposing forces or substances."[62]) As they behaved as bodies in other respects, Berzelius found it most appropriate to treat them as such. Nor were these imponderables otherwise entirely unrelated. Between heat and electricity, in particular, but also between these and light, there seemed to be some sort of elusive identity: "Where one of these is, the other is also manifested, though it is impossible for one to say where it comes from."

And suddenly Berzelius expanded his reasoning into a cosmic vision: With its light the sun exerts a decisive influence on earth. One can form a picture of these conditions with a large electrical pile that is discharged by bringing two platinum tips together. In this experiment we see precisely the connection between electricity, the cause, and light and heat, the effects of our action, for around the tips there flames a miniature sun that develops a strong heat. Surely, said Berzelius, the radiation of the real sun is a result of a similar electrical discharge, perpetually maintained by a continuous circular process.[63] Because electricity, according to Berzelius, is also the active factor in all chemical reactions, we sense the power of his vision. As electricity and heat lack gravitation toward the earth, they strive, once liberated from the chemical affinities of matter, where they otherwise act, to disperse evenly throughout the universe. What if from there they are brought out of equilibrium and attracted toward the sun!

61 Berzelius, *Lärbok i kemien*, Vol. II, pp. 553f.   62 Ibid., Vol. I, pp. 129.
63 Ibid., Vol. II, pp. 528ff.

If in a manner unknown to us this equilibrium is disturbed by the suns, if the united electricities are reflected by them like light towards gravitating masses hovering in space, on which (when the movement of the light is checked) they manifest themselves as heat, and (during the interim before they can once again be dispersed into the immeasurable) they maintain the activity of affinity both in inorganic and organic nature, then we would have some idea of the possibility of the perpetual emanation of a body from the sun, without any diminution of its mass and without the unfathomable velocity in the motion of light being able to cause any of the effects of a falling ponderable body.[64]

When Arrhenius ascribed to Berzelius a doctrine of an all-uniting, living sympathy, he could very justifiably offer the judgment he did, for he was surely thinking of these pages of the *Lärbok*. Nor should we overlook them, for they are a part of our history.

64 Ibid., pp. 555f.

4

Berzelius, Dalton, and the Chemical Atom

ANDERS LUNDGREN

Introduction

Berzelius's place in the history of chemistry has typically been traced to his early acceptance and experimental elaboration of John Dalton's atomic theory. This point of view is evident in the classical, three-volume biography of Berzelius (1929–1931) published by Henrik Gustaf Söderbaum and it has characterized many succeeding works, such as those of J. Eric Jorpes and Henry Leicester.[1] However, this view of Berzelius as Dalton's champion does not do justice to the complex nature of Berzelius's debts to his predecessors and his views of atomism. Especially it passes too quickly over the significance of Berzelius's initially very harsh criticism of Dalton's theory, which cannot be treated as an uninteresting contingency.[2] A closer look at those complexities will allow for a better understanding of Berzelius's chemistry as a whole, throw into relief the nature of Dalton's achievements, and illuminate the implications for the nineteenth century of the revolution centering on Antoine Laurent Lavoisier.

Berzelius learned Lavoisian chemistry as a student.[3] What he found in the new chemistry had much in common with the Swedish chemical tradition as represented by Torbern Bergman. Though both Bergman and

An earlier version of this paper was read at the Third National Conference on the History of Chemistry in Cosenza, Italy, 18–20, September, 1989. I thank Professor Ferdinando Abbri for making it possible for me to participate in the conference.

1 Söderbaum, *Levnadsteckning*; J. Erik Jorpes, *Jac. Berzelius* (Stockholm, 1960); Henry Leicester, *The Historical Background of Chemistry* (New York, 1956).
2 Cf. Söderbaum, *Levnadsteckning*, Vol. I, pp. 479, 534f.
3 For a study of Berzelius's relations to and development of Lavoisian chemistry, see Evan M. Melhado, *Jacob Berzelius: The Emergence of His Chemical System*, Lychnos-Bibliotek, Vol. 34 (Stockholm and Madison, 1981). Standard accounts perpetuate the myth that Berzelius was self-educated in an environment totally hostile toward all science; see Anders Lundgren, *Berzelius och den kemiska atomteorin* (Uppsala, 1979), Ch. I; and Anders Lundgren, "The New Chemistry in Sweden: The Debate That Wasn't," *Osiris*, 1988, 4:146–168.

Lavoisier were committed to certain theoretical motifs, they avoided flights of speculation and attempted to keep theoretical developments in close proximity to empirical observations. Moreover, both figures were powerfully influenced by physics, which conferred on their work, inter alia, an instrumental approach to theorizing. The chemistry taught at Uppsala after Bergman was antiphlogistic, descriptive, heavily informed by mineralogical practice, and modest in its theoretical articulation. Tacitly, content and methods of Lavoisian chemistry were taken for granted.

Chemical and Physical Atoms

Two broad kinds of atomism were current in the eighteenth and nineteenth centuries, chemical and physical, differing with respect to how they were used. The physical atom had traditionally been invoked to explain phenomena common to all forms of matter. Occurring in one, two, and occasionally three or more types (which in turn might be replaced by spiritual principles and forces), they constituted the building blocks of the next level of complication, that is, the chemical atoms. The chemical atoms in their turn gave rise through combination to more complex structures, the compound atoms. Chemical atoms could be invoked to describe unique kinds and features of matter. The combination of physical instrumentalism and descriptive chemistry in Berzelius's background gave him a great predilection for chemical over physical atoms.

Chemical atoms had played a role in characterizing and specifying unique phenomena, and they were directly related to empirical chemical findings in the laboratory, as when acidity was explained by the sharp form of the atoms. Such relations, however, became increasingly difficult to elaborate, and, in view of the constrained approach to theorizing that Berzelius inherited, it is not surprising that he lacked interest in elaborating the structure and properties of the chemical atoms. However, the common sense of a practicing chemist found it necessary to assume that atoms exist. These are called here "commonsense" atoms, as all scientists, irrespective of philosophical standpoint, needed something to signify the smallest object that common sense tells us could be reached through mechanical division of different chemical substances.[4] The chemist around 1800, then, was not interested in physical atoms (because universal properties were of little interest in a science whose traditional object was specific description). Instead, he relied primarily on the commonsense atom,

---

4 Ritter also used the word atom in phrases like "wo ist eine Sonne, wo ist eine Atome, die nicht Theil wäre, der nicht gehört zu diesem *organischen ALL*" (*Beweiss*, p. 171). The term atom here carries neither philosophical nor scientific meaning, but exemplifies its common use to designate something very small.

though he occasionally conflated it with the more explicitly articulated notion of the chemical atom.

Berzelius's predilection for the commonsense atom allowed him to make use of the writings of Johann Wilhelm Ritter (1776–1810), for whose Romanticism Berzelius possessed little sympathy.[5] In his *Treatise of Galvanism* of 1803,[6] Berzelius cited Ritter as a "man with his peculiar but great genius, [who has done] many important experiments on galvanism."[7] Berzelius made no mention of Ritter's Romanticism, however, citing only those aspects of Ritter's work that referred to specifics and lay on the empirical level. When discussing the electrolysis of water, Ritter spoke of "Wasseratom" and "Atom Hydrogen," thereby amalgamating the commonsense atom with the chemical atom. As a thoroughly antiatomistic physicist, when working with practical chemistry, Ritter could use an atomic concept without problems. Even in the same person, thus, the chemist and the physicist needed, and used, different terminology on the structure of matter.[8] When Ritter as a chemist spoke about "Atomen," Berzelius, also as a chemist, could follow him; but when Ritter, taking up the physical atom, spoke in the fascinating language of the German Romantics about "Raumerfüllungsindividuen," Berzelius doubtless shuddered and read on. The heritage of Bergman and Lavoisier made this eclecticism possible.

The Technical Background

Another neglected aspect hovering in the background of Berzelius's chemistry is the handicraft and technological tradition. This aspect has been neglected for many reasons, perhaps most obviously because historians have concentrated on the theoretical development of science. Another reason is that chemical techniques have not been as spectacular as physical ones and the instruments not as beautiful and expensive.[9] The

5 On his later aversion, see Berzelius, *Reseanteckningar*, ed. H. G. Söderbaum (Stockholm, 1903), pp. 307–309; also see Sten Lindroth and Alan Rocke in this volume, Chapters 1 and 5, respectively. Note however the slightly different position of Gunnar Eriksson, also in this volume, Chapter 3.

6 Published in Swedish as *Afhandling om glavanismen* (Stockholm, 1803).

7 *Afhandling*, p. 45. The work Berzelius referred to was, of course, *Beweis, dass ein beständinger Galvanismus den Lebensprocess in dem Thierreich begleite* (Weimar, 1798).

8 "Versuche und Bemerkungen über den Galvanismus der Volatischen Batterie," *Annalen der Physik*, 1801, 9:265–352. A similar difference between the chemist's view and the physicist's can be found with Humphry Davy; cf. M. B. Hall, "The Early History of the Concept of Element," in *John Dalton and the Progress of Science*, ed. D. S. L. Cardwell (Manchester, 1968), p. 35.

9 Books with titles like *History of Scientific Instruments* rarely treat chemical apparatus. For an overview on what little has been written, see R. G. W. Anderson, "Instruments

chemist's way through history is paved with broken glass and soot. There are simply very few instruments left to study. The difficulties are almost insurmountable when it comes to the study of what was one of the chemist's foremost techniques, smell and taste. All these techniques were important during practical work in the laboratory, both for production of chemical substances and not the least for identification, determination of composition, and hence for theoretical development. The technical background was of course also deeply connected with commercial interests, as in mining and agriculture and in the production of saltpeter and gunpowder. Industries provided raw materials and the pharmacist provided chemicals to the chemists' laboratories.

The technical background explains much of the strength in the commonsense argument. The attitude of a craftsman or a technician toward a practical problem to be solved is often not very sophisticated from a philosophical point of view, but most congenial for solving the actual problem. The worker wants to solve the problem in the simplest way, which could be discerned by common sense, from what was intuitively considered correct and as corresponding to what was seen in day-to-day work. The philosophical or scientific principle behind, for example, the trajectory of a cannon ball was of no interest when it came to bombarding a fortification in front of an army, and a miner in the beginning of the eighteenth century did not need chemistry to see if a mineral contained valuable substances.[10] Even if chemistry grew increasingly theoretical during the eighteenth century, traces of the technician's attitude can be seen when a chemist says he intuitively knows what an atom is (meaning a commonsense atom), but does not intend to plunge deeper into the philosophical aspects of the question – luckily so for the progress of science, one might add.

Another side of the problem is that during the eighteenth century there often was a tension between physicists and chemists, even a hostility, reflecting a difference between a science close to philosophy and another close to handicraft and technology. The chemist could accuse the physicist of not daring to dirty his hands or to go down into a mine. And even as this tension slowly eased during the century, the differences did not totally disappear. The story of Berzelius's relations to the atomic theory can best be described in the light of the tension between Lavoisian meth-

and Apparatus," in *Recent Developments in the History of Chemistry*, ed. Colin A. Russell (London, 1985), pp. 217–237.

10 Anders Lundgren, "The Changing Role of Numbers in 18th-Century Chemistry," in *The Quantifying Spirit in the Eighteenth Century*, eds. Tore Frängsmyr et al. (Berkeley: University of California Press, 1990), pp. 245–266.

odology and the use, emerging from the technical tradition, of a commonsense atom.

From Affinity Studies to Laws of Proportions

Eighteenth-Century Attempts to Explain Chemical Affinity

Berzelius's interest in Ritter's work and in galvanism was of course connected with his interest in chemical affinity. His reaction to Dalton must be related to his affinity studies.

Attempts to apply the Newtonian law of attraction when measuring chemical affinity failed at the end of the eighteenth century.[11] Chemists sought other ways to measure affinity. Examples cited by Berzelius include Carl Friedrich Wenzel, Richard Kirwan, and Jeremias Benjamin Richter. As physicists and philosophers, they had definite views on the structure of matter; but their measurements of chemical affinity were carried out at the specific level, and their concrete results could be used by other scientists, irrespective of differing views of matter theories.

Especially important were the attempts to measure affinity by weight measurements, that is, through quantitative determination of chemical composition, like Richter's calculations of the quantities of different acids needed to neutralize a given amount of a base (and vice versa). Of course, weight measurements in principle had been of interest to the chemists who had studied affinity from a Newtonian standpoint, but when Newtonian affinity theory was applied to chemistry, not only weight but also the shape of the assumed smallest particles had to be considered. However, the shape of the atoms remained inaccessible, and the particle concept so essential in Newtonian chemistry was stripped of any practical meaning.[12]

So when chemists focused on weight measurements and quantitative determination of chemical compounds, particles (including the commonsense atom) had ceased to play a theoretical role in chemistry. That affinity was an instance of Newtonian attraction became an empty claim, and chemical affinity was studied without any obvious references to theories

11 Arnold Thackray, *Atoms and Powers: An Essay on Newtonian Matter-Theory and the Development of Chemistry* (Cambridge, MA, 1970), is the classical study. The significance of affinity studies for the development of postrevolutionary chemistry has not been much discussed. One noteworthy exception is Henry Guerlac's work on the possible influence of Richter on Dalton, "Some Daltonian Doubts," *Isis*, 1961, 52:544–554. Even if Guerlac fails to prove his thesis, he does suggest links among affinity, proportions, and atoms that have so far been little studied.

12 As in the chemistry of Stahl and Bergman. Morveau's attempt to calculate affinity from a supposed form of the atoms did not find any followers; W. A. Smeaton, "Guyton de Morveau and Chemical Affinity," *Ambix*, 1963, 11:55–64, on p. 60.

on the structure of matter.[13] Experimental studies of affinity were carried out in the descriptive, atheoretical manner already a characteristic of affinity tables. The weight measurements took place in what can be called "a theoretical vacuum," which suited Lavoisian methodology well. Dalton's theory became one way to fill this vacuum, by making it possible to correlate weight measurements to each other. Because Dalton invoked physical atoms, the vacuum was not filled without problems for chemistry.

Berzelius and the Laws of Chemical Proportions

Berzelius's work on chemical affinity resulted in his fundamental laws of proportions.[14] Berzelius saw one root of these laws in Bergman's affinity studies, especially Bergman's work on the phlogiston content in metals, the result of which Berzelius, as he put it in 1811, "translated" into modern chemistry: "A given amount of an acid does always unite with the same quantity of oxygen in the different metal oxides, which saturate the acid."[15] Affinity studies were naturally related to weight measurements, a prominent part of late eighteenth-century chemistry, especially as developed by Bergman.[16] Berzelius also accepted Joseph Louis Proust's opinion on definite proportions on the ground that the hypothesis had been experimentally so well verified.

Supported by an overwhelming mass of experiments, Berzelius in 1810 presented two proportion laws. The first can be called the "law of unity," and it stated that when two substances combine chemically in different proportions, one substance must always be present with one smallest unit, whereas the other substance, usually oxygen, enters with one, one and a half, two, or four units.[17] The law thus did not allow combinations containing two units of one substance with three of another, whereas one with one and a half of another was allowed. The second law, the "law of oxides," stated that the amount of oxygen in one part of a salt must be an integral multiple of the amount of oxygen in the other part of the salt (a salt consisting of a compound between an acid and a base, both con-

---

13 Arnold Thackray, "Quantified Chemistry – The Newtonian Dream," in Cardwell, ed., *John Dalton*, pp. 92–108, on pp. 104f.

14 "Försök rörande de bestämda proportioner, hvari den oorganiska naturens beståndsdelar finnas förenade," *Afhandlingar i fysik, kemi, och mineralogi*, 1810, 3:162–276.

15 "Om de bestämda proportioner," *KVA, Handlingar*, 1811, 32:169–197, on p. 178. For "Kuhnians" it is obvious that Berzelius's use of the verb "translate" causes doubts about the idea of paradigmatic incommensurability. Still, in 1818, Berzelius considered Bergman's affinity studies as a forerunner of the theory of definite proportions. Also note the similarity with Richter's formulations on acid-base saturation. Berzelius, *Lärbok i kemien*, Vol. III (Stockholm, 1818), p. 2.

16 Lundgren, "The Changing Role of Numbers."   17 "Försök," p. 162.

taining oxygen). Other compositions were not allowed. The proportion laws limited the number of possible combinations, something necessary to save the doctrine of definite proportions. If all kinds of numerical relations between the constituent parts of a compound should be allowed, there would be no definite proportions.

There is no doubt that the proportion laws were fundamental for Berzelius's chemistry. An important task then became to determine the smallest unit that could enter a chemical compound. This unit can be called the "proportion-determining unit." It could be studied and measured through quantitative analysis of what Berzelius called a "composition in minimum," a compound of one unit of a certain substance with the smallest amount, that is, one unit, of oxygen.[18] Such a compound could be made the starting point for a series of oxides, the construction of which was regulated by the proportion laws.

The Laws of Proportions in Chemical Practice

The function of the proportions laws is well exemplified in Berzelius's work on ammonium, published in 1811.[19] As ammonia (volatile alkali) could amalgamate with quicksilver, Berzelius already in 1808 thought its base was a metal, which he named "ammonium."[20] Careful analysis had shown that ammonia contained only nitrogen and hydrogen, and Berzelius concluded that ammonium was a metal composed of these two gases. But all alkalis were supposed to contain oxygen, and consequently the oxygen in ammonia must be hidden in the hydrogen and/or in the nitrogen. Therefore both gases, as ammonia, must be oxides of ammonium. Originally, Davy suggested this idea to Berzelius, who first doubted its value, but in 1811 apparently accepted it.[21] From this idea Berzelius constructed a series of assumed oxides of ammonium, starting with the composition in minimum, hydrogen, continuing with ammonia, nitrogen, the nitric oxides, the nitric acids, and ending with water.

The proportion laws determined the content of the series. For example, if the oxygen content in ammonia was taken as the starting point, several problems occurred. First, there would be no room for the ammonium oxides containing less oxygen than ammonia. Second, irregularities among the proportions in the series of ammonium oxides would appear, and the oxides would not follow the law of unity. However if a place for lower oxides was created, that is, if the oxygen content of hydrogen instead

18 Ibid., p. 164.
19 "Schreiben . . . an . . . Gilbert . . . ," *Annalen der Physik,* 1811, 37:208–220, on pp. 212f.
20 "Alkaliernas decomposition," KVA, *EA,* April 1808, 6:114–121.
21 Lundgren, *Berzelius,* p. 79.

was taken as the starting point, these irregularities disappeared. The law of unity demanded that hydrogen should contain oxygen. Berzelius also assumed that there still existed unknown oxides of ammonium, needed to fill the series completely and to avoid a disturbing leap in the regular flow of proportions. The ammonium series is a striking example of how Berzelius, having established a law, gave it strong regulative power. His belief in the existence of ammonium was so strong that he once declared he had seen it.[22]

Also the law of oxides determined interpretation of experiments. In studies of the lead salts of the nitrogen oxyacids of 1812, Berzelius observed that if nitric acid is composed of ammonium and oxygen, the salts did not follow the law of oxides.[23] If, however, the oxygen in nitrogen was taken to be firmly attached to ammonium, it should not be counted, and the law of oxides was obeyed. Still, the causes for chemical proportions were not known. Berzelius could not say if multiple proportions "follow a rule, common for all substances, or depend on specific rules for each substance."[24] This is an example of the theoretical vacuum of affinity studies. But Berzelius was not satisfied with that, "for all these circumstances there must be some ground, which I hope to discover by a careful examination of several substances."[25]

It was in the middle of these studies that Berzelius encountered Dalton's theory. Initially it seemed to provide an excellent answer, but it was based on a physical atomic theory, alien to Berzelius's chemistry, both for methodological reasons (as part of the Bergman-Lavoisier heritage) and for practical reasons (particles were experimentally inaccessible). The study of proportions had been independent of any theory of the structure of matter. This circumstance was to color Berzelius's interpretation and use of Dalton.

Berzelius and Dalton

Dalton's Atomic Theory

If an attitude of disinterest or skepticism toward atoms was inherent, although not outspoken, in the science Berzelius learned, that was not

22 Berzelius and Magnus Martin Pontin, "Försök med alkaliernas och jordarternas sönderdelning," KVA, EA, May 1808, 6:110–130, on pp. 126f. A German version appeared as "Electrisch-chemische Versuche mit der Zerlegung der Alkalien und Erden," *Annalen der Physik*, 1810, 36:247–280.

23 "Zweite Fortsetzung des Versuchs, die bestimmten und einfachen Verhältnisse aufzufinden, nach welchen die Bestandtheile der unorganischer Natur mit einander verbunden sind," *Annalen der Physik*, 1812, 40:162–208, on p. 183. Cf. Melhado, *Berzelius*, pp. 228–231.

24 "Försök," p. 162 [my translation].   25 Ibid., p. 163 [my translation].

the case for Dalton.[26] Physical studies of gases and their solubility in water had given Dalton the idea that atoms of different elements were characterized by their definite weights, unique for each element. His claim that these atoms were indestructible has appropriately been called a multielementary atomism.[27] The indestructibility of the atoms, corresponding to elements as defined by Lavoisier, was self-evident to Dalton, and he found support for it in Newton and probably in God. One cannot but be struck by how naïve and uncomplicated Dalton's belief in indestructible atoms was.[28] He seemingly had no thoughts on philosophical complications, and one might ask if naïveté is not a prerequisite for scientific progress. Definite, multiple, and reciprocal proportions were easily explained, and thanks to the perhaps most important part of Dalton's theory, his rule of simplicity,[29] it was possible to determine from the weight relations in chemical compounds a specific property for each kind of atom, its relative atomic weight.

Still, there was much opposition, and David Knight has distinguished three groups of opponents, all with the common denominator that they saw the atomic theory "as a statement of laws of chemical combination."[30] Mathematicians did not think Dalton's theory had provided a mathematical chemistry worthy of the name, unitarians believed in a unity of matter, and positivists held Dalton's atom to be an unnecessary theoretical entity, even if they could accept the theory as a heuristic device. But Dalton had claimed it for real. He saw his theory as a theory of particles, and he criticized those chemists who studied affinity without mentioning between what entities affinity worked. He complained of having read 37 pages of Berthollet without having found the word particle.[31] In Thackray's characterization of Dalton's chemistry as consisting of "careful observations, bold theorizing, and dogmatic belief,"[32] bold theorizing and dogmatic belief were the most conspicuous.

26 Arnold Thackray, *John Dalton: A Critical Assessment of His Life and Work* (Cambridge, MA, 1972); Thackray, "John Dalton," *Dictionary of Scientific Biography,* Vol. III (New York, 1971), pp. 537–547; R. Angus Smith, *Memoir of John Dalton and the History of the Atomic Theory Up to His Time* (Manchester, 1856), p. 18.

27 William Brock, "Dalton versus Prout: The Problem of Prout's Hypothesis," in Cardwell, ed., *John Dalton,* pp. 240–258.

28 Curiously, the role of Dalton's religion in his scientific development has been unstudied.

29 The rule states that if only one compound of two substances was known, the compound contained one atom of each constituent; if two compounds of the same substances were known, the second would consist of two atoms of one, one of the other, and so on.

30 David Knight, *Atoms and Elements* (London, 1967), pp. 20–21.

31 "Inquiries Concerning the Signification of the Word Particle," *Journal of Natural Philosophy, Chemistry, and the Arts,* 1811, 28:81–88.

32 Thackray, "John Dalton," p. 539.

Berzelius's Criticism of Dalton

From preserved laboratory journals it is clear that in 1806 Berzelius read the article "On the Absorption of Gases in Water and Other Liquids," in which Dalton presented the first list of atomic weights.[33] It is also clear that Berzelius did not take any notice of these weights. He saw Dalton's studies as a study in physics, on the mechanics of water solubility of gases, and not as a possible contribution to chemistry.[34]

It is well known that Dalton's theory spread in chemistry through the works of Thomas Thomson and William Wollaston. Thomson in 1807 still might have had doubts about the reality of atoms, as he described Dalton's theory as a "hypothesis," which "if . . . allowed, would mean much for chemistry.[35] But he soon overcame any doubts, and in the classical paper "On Oxalic Acid" (1808), he claimed that he empirically had proved the law of multiple proportions and, thereby, Dalton's theory.[36]

Berzelius learned the chemical significance of Dalton's theory mainly through the works of Wollaston. But Wollaston's attitude toward the atomic theory was complex, and he changed his views over time in a way difficult to follow.[37] He sometimes used Dalton's theory to explain chemical proportions and could therefore be considered an atomist. On the other hand, he more often preferred the more neutral (at least from the standpoint of matter theory) "equivalent."[38] Having read Wollaston's "On the Super-Acid and Sub-Acid Salts,"[39] Berzelius interpreted Dalton's theory as a theory on the laws of chemical proportions, and in 1810 he placed great expectations on it because of this possibility. Wollaston's paper, he observed,[40] was motivated by

> Dalton's hypothesis, that when substances can be united in different proportions, these proportions are always a simple multiple by 1, 2, 3, 4 . . . , of the weight of one of the two substances, which Wollaston experimentally seems to have confirmed. This manner of looking upon chemical compounds even at a first sight throws

33 Originally published in *Memoirs of the Literary and Philosophical Society of Manchester*, 1805, 1:271–287. Berzelius read it in *Journal of Natural Philosophy, Chemistry, and the Arts*, 1806, 13:291–300.
34 KVA, MS Berzelius 24:1a; cf. Lundgren, *Berzelius*, pp. 90f.
35 Lundgren, *Berzelius*, p. 87.   36 *Philosophical Transactions*, 1808, 98:63–95.
37 D. C. Goodman, "Wollaston and the Atomic Theory of Dalton," *Historical Studies in the Physical Sciences*, 1969, 1:37–59.
38 But cf. the discussion of Wollaston in Alan Rocke, *Chemical Atomism in the Nineteenth Century* (Columbus, 1984), pp. 61–66.
39 "On Oxalic Acid," *Philosophical Transactions*, 1808, 98:96–102.
40 Försök," pp. 163f.

so much light on affinity-theory, that if Dalton's hypothesis is found to be correct, it is the greatest step for chemistry in its perfection as a science.

There are no references to atoms in Berzelius's presentation of Dalton's theory, even if he at once noticed that the theory contained a mechanical explanation of proportions. The kernel of Dalton's theory, the rule of simplicity, is discussed in terms of weight relations. This was to be expected, as theories on matter structure had had no bearing on the formulation of the proportion laws. They were the outcome of a wish to measure affinity, something earlier done without reference to particles of any kind. Consequently Dalton's theory represented to Berzelius a new way to understand affinity, not matter.

How Dalton had carried out his experiments Berzelius did not know until the spring of 1812, when he finally had a chance to read Dalton's *A New System*. He reacted as most chemists, negatively. He did not react against the concept "atom" in itself. It was familiar to Berzelius as a "commonsense" concept, meaning a very small part of matter (more or less in Ritter's sense).[41] None objected to such a use, but Dalton's theory was more, and Berzelius's statement that Dalton "a trompé mes espérances" has often been quoted.[42] But why the negative reaction? Berzelius did not raise any philosophical objections of an ontological or metaphysical type. His objections were founded in the laboratory. Besides the unsatisfying experimental verification, Berzelius at once noted that Dalton's theory was incompatible with the empirically established proportion laws. He directly accused Dalton of trying "modeler la nature d'après son hypothèse."[43] This was a deadly sin to Berzelius (which is not the same as to say that he did not sin himself sometimes, as with the ammonium series). The hope Berzelius originally had attached to Dalton's theory, that it could explain the theory of proportions, was not fulfilled: "I expected to find out whether Dalton, struggling with the same difficulties I faced, had not perhaps been more successful in overcoming them."[44] Not to have found the problems solved was a disappointment.

At the same time Berzelius preserved some of the initial enthusiasm he had felt after having read Wollaston, and he declared a future for the theory. Its simplicity was both beautiful and tempting, and it appealed to Berzelius as a possibility. But the important methodological differences

---

41 For example, in *Föreläsningar i djurkemien*, Vol. II (Stockholm, 1808), p. 484, where Berzelius speaks about "the smallest atom" of a "poison of disease" [*sjukdomsgift*].
42 Originally in a letter to the French chemist and translator H. F. Gaultier de Claubry, 27 April 1812; *Bref*, Vol. III:ii, p. 104.
43 Berzelius to Gaultier de Claubry, ibid., p. 105.
44 "Zwei Schreiben ... an ... Gilbert," *Annalen der Physik*, 1812, 42:276–298, on p. 298 [my translation].

between Dalton and Berzelius made it impossible for him to accept it. Dalton had started with indestructible atoms and explained proportions, while Berzelius had started with proportions and sought an explanation. Dalton's theory could not give that explanation.

In the summer of 1812 Dalton and Berzelius exchanged a few letters.[45] Berzelius first wrote to Dalton "il y a plusieurs points où nos resultats ne sont point parfaitement d'accord." Dalton flattened these experimental differences out, although to Berzelius they contained severe criticism, and interpreted Berzelius's result as a confirmation of his own theory. "We are agreed that an *ultimate portion* of sulphur, for instance, is nearly twice the weight of one oxygen. . . . My numbers for the *atoms* of sulphur and oxygen are 13 and 7 respectively," and Dalton rhetorically states that "the doctrine of definite proportions appears to me *mysterious* unless we adopt the atomic hypothesis." To him, definite proportions and the atomic theory were firmly linked. Certainly, answered Berzelius, Dalton could explain definite and multiple proportions; they were mysterious without the atomic theory. But he could still not accept it in its present form, even if much chemistry, including his own work, seemed to verify it.[46]

At first glance Dalton's theory seemed to have given the commonsense chemical atom an important place in chemical theory, but at the same time this atom had become endowed with physical properties, which contradicted experimentally established chemical results. The proportion laws demanded from Dalton half atoms, which of course were an impossibility. It is true that Berzelius in his work on ammonium had suggested a method that could exclude half proportions. It could be done by speculating about the existence of an oxide with a lower content of oxygen than the lowest so far known. If such an oxide existed, the relationship $1:1\frac{1}{2}$ between the oxygen content in the two lowest oxides could be replaced with a series of three oxides with the oxygen relations $1:2:3$.[47] Another method, suggested in the letter to Dalton, was to regard, for example, minium not as Pb_2O_3, but as a mixture of PbO and PbO_2.

Berzelius's readiness to propose methods to eliminate half proportions might be interpreted as a strong interest in justifying Dalton's theory. But when Dalton amalgamated the specific chemical and the universal physical atoms, he made the chemical atoms unsplittable, which made it impossible to use them for explaining chemical proportions. The proportion laws were not linked with atoms, but with the proportion-creating

45 Henry E. Roscoe and Arthur Harden, *A New View of the Origin of Dalton's Atomic Theory* (London, 1896), pp. 156–162. All quotations from these pages [emphasis added].
46 Ibid., pp. 161f.
47 Berzelius to Berthollet, 12 December 1811, *Bref,* Vol. I:i, p. 33. The example in the letter is tin oxides.

unit, which carried no distinct physical meaning. The atomic theory was subordinated to the theory of proportions. If Berzelius was to accept Dalton's theory, the theory had to allow half atoms in chemical compounds, which of course it could not. Dalton's atoms were by definition impossible to split.

Turbulence After Dalton

How to Name Proportion Units

After having read Dalton, Berzelius's thinking seems to have been in a state of turbulence. His terminology concerning proportions varied greatly, likely because of a certain ambivalence toward matter theory in general and to Dalton's theory in particular. The gap between atoms and proportions naturally cast a shadow on atoms, and Berzelius preferred to express proportions in weight relations.

In 1813, Berzelius with his friend Alexandre Marcet published an article on sulfuret of carbon, with an appendix written by Berzelius.[48] In the appendix he sought support for his weight determinations, from both Humphry Davy (who certainly was not a Daltonian) and Dalton. Dalton had given the relation between "the weights of [carbon and sulfur] particles" as 5:13, Davy the proportions as 11.4:30, and "either of these modes of computation correspond very nearly to two atoms or portions of sulphur to one of carbon . . . which is perfectly agreeable to the doctrine of definite proportion." Similarly in the appendix he observed that "Sir *Davy* . . . adopts, like Mr. *Dalton*, the idea that sulphurous acid gas consists of one portion of sulphur to two of oxygen."[49] The atomic theory was always subordinated to the proportion theory; "the supposed atom, or ultimate particle, *or that which is to be regarded as the single portion of a body,* requires to be fixed according to some determinate law, before Mr. Dalton's ingenious *method of expressing* the proportions in which bodies combine, can be applied with certainty and precision."[50] When Berzelius used the concept atom, it was above all to express a certain weight unit, found when measuring proportions.[51] The simplicity and the future possibilities of the theory remained appealing,

48 Berzelius and Alexandre Marcet, "Experiments on the Alcohol of Sulphur, or Sulphuret of Carbon," *Philosophical Transactions,* 1813, *103*:171–199; pp. 188–199 comprise the appendix.
49 Ibid., pp. 187, 191. The German translation has "Verhältnisstheile"; cf. "Versuche über den Schwefelalkohol," *Journal für Chemie und Physik,* 1813, *9*:284–300, on p. 295.
50 Berzelius and Marcet, "Experiments on the Alcohol of Sulphur," pp. 191f; emphasis added.
51 For other examples, see Lundgren, *Berzelius,* p. 130.

and therefore he did not always clearly distinguish "atom" (meaning weight unit) from "atom" (meaning a particle and corresponding to the chemical atom).

That Berzelius did not wholeheartedly accept Dalton's theory is clear from his readiness to express proportions in other ways than with atoms. One alternative was provided by the volume theory, formulated by Joseph Louis Gay-Lussac and originally stating that when two gases chemically react, the proportions between the volume of the initial gases and that of the product can be expressed in small whole numbers. Berzelius learned about Gay-Lussac's work around 1810, when he knew about Dalton but had not yet read him in the original. In spite of the disagreement between Dalton and Gay-Lussac, to Berzelius both confirmed the theory of proportions.[52] Berzelius said he saw "a wonderful agreement" between Gay-Lussac's law and Dalton's hypothesis.[53] The reason for the agreement was that relations between proportions in both cases could be expressed in small integers.

After the disappointment with Dalton, Berzelius often (but not consistently) used "volume" to express proportions. For example, in 1813, Berzelius gave the composition of nitric acid in the following manner "1 volume (atom or proportion) of nitric and 4 volumes of oxygene."[54] Volumes and atoms were both proportion weight units, but the decisive advantages with volumes were their firmer experimental foundation and their providing an alternative (the half volume) to a half atom. However, because the proportion laws were based neither on atoms nor on volumes but on the proportion-creating unit, a change of terminology did not mean much for the practical chemical result. When Berzelius in 1813 gave what looks like a definition of atoms, it was not surprisingly in terms of weights: "I would consider as the atom of any other gas [besides oxygen] the proportional weight of an equal measure of that gas [e.g., oxygen].[55] The "definition," or perhaps better, "explanation," of an atom does not correspond to a Daltonian atom.

During the turbulence created by Dalton's theory, the basic problem still remained that the atomic theory gave to matter a physical property that contradicted the chemical theory of proportions. The two theories partly overlapped, as the atoms gave an opportunity to explain chemical

52 M. Crosland, "The Origin of Gay-Lussac's Law of Combining Volumes of Gases," *Annals of Science,* 1961, 17:1–26. For the relations between Dalton and Gay-Lussac, see Rocke, *Chemical Atomism in the Nineteenth Century,* pp. 40–42.
53 *Lärbok i kemien,* Vol. II (Stockholm, 1812), pp. 575f.
54 Berzelius, "Experiments on the Nature of Azote," *Annals of Philosophy,* 1813, 2:276–284, 357–368, on p. 364. Cf. also letter to Gahn, 13 January 1814: "or, after Dalton, atom" [my translation]; *Bref,* Vol. IV:ii, p. 95.
55 Berzelius and Marcet, "Experiments on the Alcohol of Sulphur," p. 192n.

proportions, but because the two theories were not congruent problems occurred. Certainly Dalton had created "a corpuscular theory from which the doctrine of definite proportions follows as a necessary consequence,"[56] but to Berzelius the reverse was not necessary. From the existence of proportions the atomic theory does not follow as a logical consequence.

Berzelius, however, believed there was a possibility that the atomic theory could explain proportions. In late 1813, in what can be called a summary of his views, "Essay on the Cause of Chemical Proportions,"[57] he credited Dalton as being the first to have tried such an explanation. Berzelius now more consciously used the "commonsense" argument for accepting chemical atoms and for taking Dalton seriously. That matter was built by unsplittable atoms was simply "most conformable to our experience." This vague expression, however, did not solve the basic problem. Furthermore by allowing $2A + 3B$, Dalton also had allowed $9A + 10B$, $99A + 100B$, and so on, which destroyed the experimentally well-established proportion theory. Berzelius had saved definite proportions by the limits inherent in the proportions laws, but this approach entailed the necessity of half atoms.

Some of the problems could easily be solved. According to the law of unity the composition $1:1\frac{1}{2}$ was possible, which was contrary to Dalton. But as has been shown the half proportion (and hence atom) could be made to disappear by supposing hitherto unknown combinations from which to start the series of multiple proportions.[58] The law of oxides offered greater problems and still contained "great difficulties" when applied "to a number of chemical phenomena."[59] If the law of oxides was to be obeyed, one part of the salt had to have the formula $1A + 1\frac{1}{2}B$, something Dalton could not allow. Dalton, on the other hand, would write $2A + 3B$, which was impossible to Berzelius.[60] Before this problem was solved "the hypothesis of atoms can neither be adopted, nor considered as true."[61] "Hypothesis of atoms" here stands both for the use of chemical atoms to explain proportions and for a matter theory. The problem with the half atom could have been solved with the volume theory, but its earlier strength had now become its weakness. Its experimental support was limited to gaseous substances.

"Cause of Proportions" shows an ambivalent Berzelius. He apparently liked the idea of explaining proportions with particles, and he expressed

56 "Experiments on the Nature of Azote," pp. 277f.
57 Berzelius, "Cause of Proportions" (see List of Abbreviations).
58 "Schreiben an Gilbert," pp. 212f.   59 Berzelius, "Cause of Proportions," p. 445.
60 For a full discussion, see Colin A. Russell, "Berzelius and the Development of the Atomic Theory," in Cardwell, ed., *John Dalton*, pp. 259–273, on p. 267.
61 Berzelius, "Cause of Proportions," p. 450.

himself as if accepting atoms, even if the physical consequences of this assumption were at odds with the chemical theory of proportions. Berzelius's varying terminology reflects the fact that the commonsense atom, however much it would simplify chemistry, still was in conflict with experimentally determined proportions. How to name proportions was determined by what kind of experimental support had been found for the theory of proportions or by what chemist had been supporting it.

Berzelius and Particles

Though the laws of proportions were in principle independent of physical theories, in several respects, apart from the theory of proportions, Berzelius's chemistry drew strength from conceptions of particles, even physical ones. One example is Berzelius's explanations of affinity. The laws of proportions having emerged as the outcome of attempts to measure affinity, Berzelius wanted an explanation of affinity itself, something that Dalton's chemistry could not provide. To this end, Berzelius invoked a kind of electrical polarization, which, in turn demanded a particle as the locus of polarization. "It seems probable to me that the specific electrochemical nature of bodies must lie in an electrical polarity of the integrant particles."[62] Though he thus introduced a physical element into his thinking, the linkage it achieved between affinity effects and the laws of proportion doubtless granted its legitimacy in Berzelius's eyes.

In other instances, Berzelius invoked a more commonsense atom to help answer the question whether there is an upper limit on the number of proportion units of one substance that can be attached to one unit of another substance. Supposing that all atoms, irrespective of relative weight, have the same size, Berzelius noted that for geometrical reasons only 12 atoms of one kind could be packed around one of another. Accordingly, 12 was the upper limit.[63]

This commonsense argument might have been reinforced by the common use of a particle concept in crystallography, where crystal form by tradition was explained in a mechanical way by using particles as "bricks."[64] The argument had no direct influence on Berzelius's chemis-

---

62 In a letter to Schweigger, 4 December 1813, printed as a footnote to the German translation of "Experiments on the Alcohol of Sulphur," *Journal für Chemie und Physik,* 1813, 9:296f. [my translation]. On Berzelius and electricity, see Colin A. Russell, "The Electrochemical Theory of Berzelius," *Annals of Science,* 1963, 19:117–145; Trevor Levere, *Affinity and Matter* (Oxford, 1971), Ch. 5.

63 Berzelius, "Cause of Proportions," p. 449.

64 The parallel with Wollaston is instructive. Despite his often uttered skepticism toward Dalton, Wollaston used particles when discussing crystals; see, for example, "The Bakerian Lecture, on the Elementary Particles of Certain Crystals," *Philosophical Transactions,* 1813, 103:51–63.

try, as it was difficult to correlate chemical and crystallographic properties, but it contributed to an awareness of the usefulness of a particle concept.

This sort of utility that Berzelius drew from atoms and the hope that some version of an atomic theory might serve to explain proportions conspired to maintain the turbulence evident in the "Cause of Proportions." For example, in his mineralogy of 1814, Berzelius conveys a clear sense of support for atoms.[65] However, his uncertainty about the appropriate language for describing the proportion-determining unit persisted, as is suggested by his close juxtaposition of the phrases, "volumes (atoms, particles)" and "particles (atom, volume),"[66] as well as by his persistence in adding phrases like "atoms, according to Dalton," when describing proportions. Or in phrases like "Mr. Dalton a etabli une *autre manière de envisager* les proportions determinés, manière qui se distingue par son extrême simplicité," according to which sulfuret of carbon "est composé de 2 molecules (particles) de soufre."[67] And again, "atom, according to Dalton's way of expression, volume according to mine."[68]

The principal obstacles for the fusion of the commonsense atom with the proportion-determining unit lay in the half proportion demanded by the law of unity and the law of oxides. Concerning chemical proportions, the particulate structure of matter was not as self-evident, as in the case of electricity, where no conflicts with experience were seen. Experimental results in combinations with the laws of proportion demanded half particles. As noted earlier, the problem posed by the law of unity could be solved by finding lower oxides that would render the half proportions merely apparent. The problem posed by the law of oxides was not solved until 1816, in a study of phosphoric acid and its salts.[69] The phosphates could be made to agree with the law only if phosphorus contained oxygen, something Berzelius could not verify experimentally. The verdict of the experiments led to a change in the theory. From now on compounds of the type $2A + 3B$ were permitted, contrary to the proportions laws

---

65 *Försök att genom användandet af den electrokemiska theorien och de kemiska proportionerna grundlägga ett rent vettenskapligt system för mineralogin* (Stockholm, 1814) *An Attempt to Establish a Pure Scientific System of Mineralogy by the Application of the Electro-Chemical Theory and the Chemical Proportions*, tr. John Black (London, 1814). Here quoted from the English translation.

66 Ibid., pp. 56f.

67 KVA, MS 27:8a fol 8; emphasis added. From the preserved manuscript (in French) to the article on sulfuret of carbon. The contemporary English translation of this has "atoms" for "molecules (particles)."

68 Berzelius to Marcet, 22 April 1813; *Bref*, Vol. I:iii, pp. 36–41. For more examples see Lundgren, *Berzelius*, pp. 172f.

69 "Untersuchungen über die Zusammensetzung der Phosphorsäure, der phosphorigen Säure und ihrer Salze," *Annalen der Physik*, 1816; 53:393–446, 1816 54:31–55.

(though perhaps only as occasional exceptions). There being no half proportion, no half atom was needed. Thus, both laws had moved from
contra to pro atoms. From now on, Berzelius expressed himself consistently in terms of particles, having linked the proportion-determining unit
with the chemical commonsense atom. He did not accept chemical atoms
as dogma, however, for the instrumentalism of his heritage would not
permit that.

Atomist or Not?

With so many arguments for particles could not Berzelius be said to have
been an atomist? Of course, but some specification is needed. He was not
Daltonian in a strict sense, as neither Dalton nor his fellow chemist Thomas
Thomson thought. In 1813 Thomson questioned the law of oxides, the
greatest obstacle for Berzelius to accept Dalton's theory: "This canon
appears to me at present to be entirely empyrical. I cannot see any sound
reason (different from the result of analyses) that could lead us to adopt
it."[70] To say that the proof of the law, its empirical foundation, was its
weakness, made Berzelius furious, and could hardly have changed his
opinion on the atomic theory to the better. Thomson's article was "un
chef-d'oeuvre de charlatanerie, de négligence et de manque de consé-
quence."[71] But even after Berzelius had removed this obstacle Dalton
himself did not see Berzelius as an adherent of his atomic theory. As late
as 1830 he found it remarkable that "such men as Berzelius, Davy and
even Wollaston" were not aware "that they were speculating upon the
atoms or ultimate combining particles of bodies."[72]

   In the third volume of his textbook (1818), the bulk of which, published in French translation, became well known as his *Essai sur la théorie des proportions chimiques,* Berzelius explicitly adopted a conscious,
instrumentalist methodology for the use of scientific theories generally
and the atomic theory in particular.[73] The methodology took form during the turbulence that followed Berzelius's encounter with Dalton,
emerging in the "Cause of Proportions," the mineralogy, and finally the
textbook. In the mineralogy, for example, Berzelius juxtaposed two approaches to the cause of proportions, the atomic explanations and the

70 "On the Daltonian Theory of Definite Proportions [first installment]," *Annals of Phi-
    losophy,* 1813, 2:32–52, on pp. 39f.
71 Letter to Marcet, 24 October 1813, *Bref,* Vol. I:iii, p. 78. But Thomson maintained his
    view that Berzelius was not an atomist; cf. Knight, *Atoms and Elements,* p. 87.
72 Thackray, *John Dalton,* pp. 94, 115; William H. Brock, "Dalton versus Prout," p. 248.
73 *Lärbok i kemien,* Vol. III (Stockholm, 1818); *Essai sur la théorie des proportions chi-
    miques* (Paris, 1819). [Translations in this paragraph and the next are mine.]

volume theory."[74] The latter was based on experience (the law of combining volumes), but the former, "notwithstanding it gives rise to difficulties which cannot immediately be solved or removed, . . . coincides best with our ordinary manner of seeing and conceiving bodies and their composition."[75] Atomism was supported by common sense and could play an instrumental role.

In 1818 the methodology was clearly formulated, and Berzelius now doubtless considered the atomic theory an instrumental theory. When discussing theories for explaining proportions, he explicitly says, "All theories are nothing but a manner of representing, in a consistent way, what is really behind the phenomena, and it is probably enough if a theory can explain all known facts. Such a manner of representation might be false, and certainly often is," even if, in a given state of science, it is the best way to represent the phenomena.[76] To explain proportions, Berzelius from 1818 chose particles and named them "atoms," as best "corresponding to our idea" of the smallest proportion unit. He called the theory a corpuscular theory, corpuscles corresponding to chemical atoms. These he instrumentally considered unsplittable. His next step was to ascribe to atoms certain properties not found by experiment, such as spherical form, uniform size, and so forth, to facilitate the mechanical explanation of affinity, the crystalline form of compound atoms and so on.[77]

He consciously refrained from discussing the metaphysical question of infinite divisibility and did not philosophize on the ultimate structure of matter.[78] His "instrumentalism" freed the chemical atoms from metaphysical content. He was never interested in ontological questions and treated them offhandedly. In 1818 it was enough if the smallest proportion-creating unit could be theoretically represented as a particle. He accepted Dalton's atoms as commonsense chemical atoms, not as philosophical physical atoms. Dalton's atoms and Berzelius's corpuscles in practice finally coincided, but the two had reached their respective standpoints along different routes. Dalton's atoms belonged to a different philosophical tradition from Berzelius's particles. An atomic theory with roots in affinity studies, for which particles at the end of the eighteenth century were of no interest, could develop more instrumentally than Dalton's atomic theory, based as it was on a dogmatic interpretation of Newtonian mechanics. The difference between the two chemists was overshadowed from 1818 by Berzelius himself, when he decided to name the particles that chemical elements were considered to consist of "atoms,"

74 *Attempt*, pp. 109–111; cf. "Cause of Proportions," pp. 444–450, and above, text accompanied by nn. 57, 65–68.
75 *Attempt*, p. 111.   76 *Lärbok i kemien*, Vol. III, p. 16.
77 Ibid., p. 21.   78 Ibid., p. 20.

as the "most common and best expression on the idea" associated with these particles.[79] Berzelius thus allowed the atomic theory to give a new theoretical significance to stochiometric relationships in chemistry.

Discussion and Conclusions

It is obvious that the Lavoisian tradition continued to exercise a strong influence on chemistry long into the nineteenth century. It is in fact surprising how little the relations between Lavoisier and Dalton have been studied, although just 20 years separate Lavoisier's *Traité de Chemie* from Dalton's *New System*. Furthermore, the relations between the atomic theory and the antecedent chemical revolution are still unclear.[80] In the literature on Dalton not much is said about Lavoisier, and vice versa, but from Berzelius's chemistry emerge many connections between these two works.[81]

First, in regard to the role of oxygen and the composition of salts, Lavoisier's assumed composition of salts was a necessary prerequisite for the law of oxides, which in turn created the greatest obstacle to Berzelius's acceptance of Dalton's theory. Second, even if Lavoisier did not formulate a proportion theory, his emphasis on quantitative analysis was also one prerequisite for the theory of definite proportions, and therefore for a fruitful amalgamation of Lavoisian chemistry with Dalton's theory. Third, the Lavoisian definition of an element had some obvious similarities with Berzelius's proportion-creating unit. In principle, both should be possible to find during laboratory work, but both were also part of a theoretical framework that transcended the empirical foundation. In Berzelius's case the more general theory of proportions made it possible theoretically to calculate from weight measurements the composition of both the unit and its compounds, which results so far did not have a counterpart in the laboratory. Fourth, with electricity Berzelius added a new substance to the important group of imponderables, central to Lavoisian chemistry.[82]

79 Ibid., p. 21. This is obviously at odds with Alan Rocke, who in his otherwise excellent study on chemical atomism in the nineteenth century has argued that Berzelius was not a "positivist"; Rocke, *Chemical Atomism in the Nineteenth Century*, p. 70.

80 This has been pointed out by J. B. Gough, "Lavoisier and the Fulfillment of the Stahlian Revolution," *Osiris*, 1988, 4:15–32; and Robert Siegfried, "The Chemical Revolution in the History of Chemistry," ibid., pp. 34–52.

81 A valuable exception is Melhado, *Berzelius*. See also Melhado, "Mineralogy and the Autonomy of Chemistry around 1800," *Lychnos*, 1990:229–262.

82 To study the continuity from the eighteenth century, a closer comparison between the French chemists and Berzelius should be of interest. On Berthollet see Maurice Crosland, "The First Reception of Dalton's Atomic Theory in France," in Cardwell, ed., *John Dalton*, pp. 274–287; Crosland, *Gay-Lussac: Scientist and Bourgeois* (Cambridge, 1978), pp. 136–140.

In 1828 Berzelius and the Swedish Academy of Sciences had the opportunity to move to a bigger building, located in the center of Stockholm. As perpetual secretary, Berzelius had his headquarters here until his death in 1848. From the archives of the Royal Swedish Academy of Sciences, used with permission.

Berzelius also in many ways exemplifies the role of chemical technology and handicraft. First must be emphasized his experimental skill, which never has been questioned, and rightly so. He constantly underlined the necessity of carefully done experiments as fundamental in chemistry, and throughout his life he actively developed chemical apparatus and experimental methods. He did glassblowing, he used balance and the blowpipe, he exploited smell and taste (his experience after smelling hydrogen selenide was terrifying).[83] To Berzelius an argument based on carefully done experiments was a very strong argument, as is obvious in the case of the laws of proportions. Second, his support for commonsense chemical atoms offers a good example of the technological attitude. Third, the lingering tension between his experimental results and the physical implications of Dalton's atomic theory can be seen as a prolongation of the conflict between chemists and physicists. Fourth, and finally, in his mineralogical work, as well as his partnership in the chemical factory of Gripsholm, there is an obvious link to commercial interests, and during his whole career Berzelius was very much occupied with what must be called applied science.

83 Söderbaum, *Levnadsteckning*, Vol. II, p. 94; Vol. III, pp. 463f.

To understand Berzelius's changing attitudes toward the atomic theory, stress must be put on the tension between the instrumentalism in the Lavoisian methodology and the commonsense atom prevailing in the chemical technological tradition. The chemists who developed the new methodology during the chemical revolution had been inspired by the successful experimental physics.[84] However, the Scandinavian and German chemists who used the methodology around 1800 were usually not trained in physics, but were trained in a chemical tradition close to the practical technologies of mining and pharmacy and imbued with common sense.

Still, it must be recalled, there was nothing in Berzelius's criticism that *directly* contradicted Dalton's opinions on the structure of matter, and no arguments directed against atoms per se, that is, against physical atoms. Such a criticism would have to be philosophical, and philosophical questions of that kind never caught Berzelius's interest. What Berzelius criticized was Dalton's use, through his views of chemical combination, of atoms to explain chemical proportions.

To Berzelius, theories were not the same as truth but "bridges to the truth."[85] He was not a philosopher but very conscious about methods. He had distinct ideas on the role and status of theories, as well as on the role and status of empirical investigations. This consciousness grew during the confrontations with Dalton's atomic theory and was to enable him to make productive use of the atomic concept. His methodological interests grew naturally out of the clash between the methodology of the Lavoisian chemistry and the commonsense atom, emanating from a blend of Swedish mineralogical tradition and lingering Bergmanian Newtonianism. In this clash the atomic concept changed. From having been a natural and self-evident commonsense concept, it became an instrumental concept in a conscious methodology and a prerequisite for the rapid development of chemistry during the nineteenth century.

84 *Osiris*, 1988, 4, contains many articles dealing with this theme.
85 *Öfversigt af djur-kemiens framsteg och närvarande tillstånd* (Stockholm, 1812), p. 9.

5

Berzelius's Animal Chemistry: From Physiology to Organic Chemistry (1805–1814)

ALAN J. ROCKE

There has been much discussion, and little agreement concerning Jacob Berzelius's philosophy of biology. Most historians of science have concluded that Berzelius was a committed vitalist,[1] a viewpoint that recently has been persuasively opposed by Bent Jørgensen and John Brooke.[2] The confusion is highlighted by the fact that even in E. Benton's sophisticated "three-dimensional typology" of nineteenth-century vitalism or in Brooke's amazing delineation of 160 possible interpretations of the statement of the impossibility of organic synthesis, Berzelius still eludes unambiguous categorization.[3] And B. S. Jørgensen's necessary distinction between chemical and physiological vitalism is not really sufficient, as Benton has noted.[4]

The research for this paper was supported in part by a sabbatical leave from Western Reserve College during the spring of 1985. For their helpful comments on earlier drafts of this paper I am grateful to John Brooke, Virginia Dawson, Tore Frängsmyr, and Evan Melhado.

1 For example, J. Jacques, "Le vitalisme et la chimie organique pendant la première moitié de XIXe siecle," *Revue d'histoire des sciences,* 1950, *3*:32–66; T. O. Lipman, "Wöhler's Preparation of Urea and the Fate of Vitalism," *Journal of Chemical Education,* 1964, *41*:452–458, on pp. 454–455; and J. R. Partington, *A History of Chemistry,* Vol. IV (London: Macmillan, 1964), p. 252.

2 B. S. Jørgensen, "Berzelius und die Lebenskraft," *Centaurus,* 1964, *10*:258–281; Jørgensen, "More on Berzelius and the Vital Force," *Journal of Chemical Education,* 1965, *42*:394–396; John H. Brooke, "Wöhler's Urea, and Its Vital Force? – A Verdict from the Chemists," *Ambix,* 1968, *15*:84–114, on pp. 85–89; Brooke, "Organic Synthesis and the Unification of Chemistry – A Reappraisal," *British Journal for the History of Science,* 1971, *5*:363–392, on pp. 380–382.

3 E. Benton, "Vitalism in Nineteenth Century Scientific Thought: A Typology and Reassessment," *Studies in the History and Philosophy of Science,* 1974, *5*:17–48, on p. 24n; Brooke, "Vital Force?" pp. 85n, 87n.

4 Jørgensen, "Lebenskraft," p. 278; Benton, "Typology," p. 19. Similarly, the distinction between "organic" and "organized" matter, that is, between organic compounds and components of living tissues, seems to have been a more important distinction to physiologists than to Berzelius; see Lipman, "Fate of Vitalism," p. 457; Brooke, "Organic

This ambivalence is not the result of careless historical analysis, but rather emerges directly from the primary sources. In his first paper devoted to elemental organic analysis (1814), Berzelius affirmed that organic compounds could not be synthesized, as "nature" impresses on them, via the nervous system, distinct new electrochemical modifications not found in the inorganic realm. He defended this view in all editions of his extraordinarily popular textbook, adding that the "*something*, which we call vital force, lies entirely outside the inorganic elements, and is not one of their original properties."[5] But in a less well known work, Berzelius also averred that all functions of the body, including thoughts and emotions in the brain, have purely physicochemical causes, and throughout his life he spurned any suggestion that the laws of inorganic nature were different in living matter.[6] The idealist Swedish physician and philosopher Israel Hwasser (1790–1860) consequently denounced Berzelius's "sordid materialism."[7]

Part of the difficulty in analyzing Berzelius's position has been the result of a prevalent impression that vitalism need necessarily contain occult or animistic elements and is inconsistent with materialism. It will be argued here that a solution to this conundrum is to be sought by viewing Berzelius in the tradition of vital materialism, a combination of vitalistic and materialistic thought.[8]

But there are broader concerns here than attempting to delineate a consistent philosophy of biology in Berzelius's works. Berzelius's earliest scientific research, from 1805, was in physiological, or "animal," chemistry,[9] and he initially thought he would remain in this field. However,

Synthesis" and "Vital Force?" p. 103. In this I concur with Wilhelm Prandtl, *Humphry Davy, J. Jacob Berzelius* (Stuttgart: Wissenschaftliche Verlagsgesellschaft, 1948), p. 213.

5 Berzelius, "Experiments to Determine the Definite Proportions in Which the Elements of Organic Nature Are Combined," *Annals of Philosophy*, 1814, 4:323–331, 401–409; 1815, 5:93–101, 174–184, 260–275; on 4:323, 329; Berzelius, *Lehrbuch der Chemie*, 2nd ed. (Dresden and Leipzig, 1825–1831), Vol. III:1 (1827), pp. 136–137; ibid., 5 ed. (1856), Vol. IV, p. 2.

6 Berzelius, *Föreläsningar i djurkemien* (Stockholm, 1806–1808), Vol. I, pp. 1–4, 78; Berzelius to C. A. Agardh, 22 November 1831, in *Bref*, Vol. IV:iii, pp. 71–72.

7 *Bref*, Vol. IV:iii, p. 153; Erik Nordenskiöld, *The History of Biology* (New York: Knopf, 1929), pp. 292–293, 373; S. E. Liedman, "Israel Hwasser," *Svenskt biografiskt lexikon*, Vol. XIX (Stockholm, 1972), pp. 513–516; H. G. Söderbaum, "Berzelius und Hwasser, ein Blatt aus der Geschichte der schwedischen Naturforschung," in J. Ruska, ed., *Studien zur Geschichte der Chemie* (Berlin; Springer, 1927), pp. 176–186. The term sordid materialism is Söderbaum's; I have not found it in Hwasser's writings.

8 Timothy Lenoir, *The Strategy of Life: Teleology and Mechanics in Nineteenth Century German Biology* (Dordrecht: Reidel, 1982), esp. pp. 17–37.

9 These terms are not, of course, interchangeable, and "physiological chemistry" was not often used in the early nineteenth century. However, my focus within Berzelius's animal chemistry is on those aspects that relate to physiology, and my focus within his physio-

his views and activities underwent a number of important modifications in the following decade. This activity culminated in a major paper on the chemistry of simple organic compounds, a landmark contribution in which we see for the first time a number of characteristics that were to mark organic chemistry for decades thereafter. It is also argued that Berzelius's early studies of animal chemistry not only led him to the study (and transformation of the content) of organic chemistry per se, but also influenced and reflected his thoughts on religious and philosophical subjects. Such a purpose cannot be achieved by treating his science alone, and so I begin with some biography.

Berzelius's father, a Lutheran pastor and schoolteacher, died when Berzelius was four years old. Two years later his mother married Anders Ekmarck, the pastor of a German congregation in a neighboring village. Ekmarck took responsibility for the education of his stepson as well as that of his own children; he performed this office with care and regard, earning Berzelius's deepest love and respect. According to those who knew him, Ekmarck was a man of exemplary virtue and no little learning, and Berzelius related that the most "earnest worship of God" always prevailed in his new home. Not only his father and stepfather, but also a number of other of Berzelius's relatives were pastors,[10] and it is not surprising that Berzelius himself planned a church career, in accordance with his mother's oft-expressed wish.[11]

There is no evidence of a crisis of faith when, at the age of 15, Berzelius abandoned the intention to pursue a church career for the study of medicine, but it seems clear that after the switch his theistic tendencies were increasingly moderated by a considerable dose of late eighteenth-century rationalism and materialism. As Söderbaum put it, Berzelius "was a child of the Gustavian period" of the Swedish Enlightenment "and was educated under the dominance of French taste."[12] This new orientation was nurtured during his years at Uppsala University between 1796 and 1802. Uppsala had had a distinguished chemical history, but both Torbern

---

logical chemistry is on those aspects that pertain to animals rather than plants. My use of the term organic chemistry is to be interpreted as indicating what Berzelius and others in his day referred to as animal and vegetable chemistry.

10 In addition to his father and stepfather, Berzelius's paternal grandfather and great grandfather, his stepmother's previous husband, and his sister's husband all were churchmen, as well as half a dozen other of his blood ancestors; see Söderbaum, *Levnadsteckning*, Vol. I, Appendix, Table 4.

11 Berzelius, *Autobiographical Notes*, tr. O. Larsell (Baltimore: Williams & Wilkins, 1934), pp. 1–7, 159; Söderbaum, *Berzelius's Werden und Wachsen, 1779–1821* (Leipzig, 1899), p. 3; J. Erik Jorpes, *Jac. Berzelius: His Life and Work* (Stockholm: Almqvist & Wiksell, 1966), pp. 13–15.

12 *Autobiographical Notes*, pp. 123–128, 179–180; Söderbaum, "Berzelius und Hwasser," p. 177.

110   *Alan J. Rocke*

Bergman and Carl Scheele had died in the previous decade, and their successors had failed to keep step with the antiphlogistic revolution that had just swept Europe. From 1798 to 1799 Berzelius learned his first chemistry from private study of a Swedish translation of A. F. Fourcroy's *Philosophie chimique* and from Christoph Girtanner's *Anfangsgründe der antiphlogistischen Chemie.*[13] He was immediately "irrevocably gripped" by experimentation, and he soon set up a makeshift laboratory in his rented digs. The professor of chemistry Johan Afzelius was a phlogistonist, whose relations with the brash and self-confident young student were somewhat uneven; the adjunct A. G. Ekeberg had been a student of Klaproth and reportedly gave excellent up-to-date lectures, but Berzelius was too poor to afford the fee. Whence it comes that Berzelius was largely self-taught in chemistry.[14]

Just as Berzelius joined the small clique of antiphlogistonists in chemistry, he also consorted with liberal and progressive circles among the students.[15] It appears that Berzelius studied his medical subjects, and especially physiology, with considerable enthusiasm. Indeed, his graduation thesis was not chemical at all, but physiological, and was presided over by Johan Afzelius's younger brother, the professor of medicine Pehr Afzelius. It is not surprising, then, that he very early settled on a field of research that combined physiology with experimental chemistry, namely, animal chemistry.[16] It is probably also relevant that his first chemical guide had been Fourcroy, who had made a considerable reputation in that area.[17]

13 Fourcroy, *Philosophia chemica eller grund-sanningar af den nya chemien,* tr. A. Sparrman (Stockholm, 1795); C. Girtanner, *Anfangsgründe der antiphlogistischen Chemie* (Berlin, 1792).
14 *Autobiographical Notes,* pp. 16–38. A. Blanck, "Berzelius som medicine studerande," *Lychnos,* 1948/1949:168–205, suggests that Berzelius was not quite the autodidact he depicted in his autobiography and that Uppsala was not as chemically backward as it may have seemed to the young antiphlogistonist. See also Anders Lundgren, "The New Chemistry in Sweden: The Debate That Wasn't" *Osiris* (n.s.), 1988, 4:146–168.
15 Blanck, "Berzelius som medicine studerande," pp. 170–176, 201–204; Söderbaum, *Werden und Wachsen,* pp. 70–74; Söderbaum, "Berzelius und Hwasser"; Evan M. Melhado, "Chemical Analysis and Chemical Atomism: Beyond the Empirical Tradition," paper delivered at the Philadelphia meeting of the American Chemical Society, 29 August 1984. Other antiphlogistic students at Uppsala at this time included Berzelius's stepbrother L. Kristoffer Ekmarck and his future assistant M. M. Pontin. Ekeberg and the professor of medicine Pehr Afzelius had published (anonymously!) the antiphlogistic *Försök till svensk nomenklatur för kemien* (Stockholm and Uppsala, 1795).
16 Blanck, "Berzelius som medicine studerande," pp. 172–175, also emphasizes Berzelius's initially strong *biological* orientation and the degree to which his future call for the application of a physiological orientation to chemistry were prefigured, and presumably at least to some extent instilled, by Pehr Afzelius.
17 *Autobiographical Notes,* pp. 27, 36–38, 61–62. Berzelius retained a strong interest in physiology for the rest of his life; see V. Kruta, "Berzelius's Interest in Physiology," *Lychnos,* 1973/1974:256–62.

First, of course, Berzelius had to begin a career, and for a variety of reasons this proved to be no easy accomplishment. Immediately on graduation he was named adjunct in medicine and pharmacy at the Stockholm School of Surgery (later the Karolinska Institute), but the position was unpaid, so he also became physician to the poor. In his meager spare time from these two occupations he was able to pursue research, but the subject, mineral analysis, was dictated by his patron and collaborator, the wealthy mine owner Wilhelm Hisinger. Despite Berzelius's pronounced distaste for inorganic chemistry at this time, the work proved successful beyond their dreams; the two Swedes concluded a brilliant investigation of electrolysis of salts that anticipated a portion of Humphry Davy's later work, and also discovered a new element, cerium.[18]

Berzelius was finally freed to pursue his own independent research by two events in the summer of 1806: the completion of a new laboratory at the surgical school, and the death of the professor of chemistry, which led a few months later to Berzelius's promotion. Later he recollected: "When I studied physiology in Uppsala . . . [it] appeared to me likely that a chemist-physiologist should be able through these sciences to accomplish a great deal, and I determined to fit myself for investigations in this direction. I therefore began to work out a chemical physiology."[19] He had been giving lectures on this subject at the school since 1803, and in the spring and summer of 1805 he wrote them up for publication. The following year they appeared as the first volume of his *Föreläsningar i djurkemien* [Lectures on animal chemistry], supplemented by a second volume two years later.[20]

The course that the young Berzelius charted for himself, physiological chemistry, was by no means an uncultivated discipline at the beginning of the nineteenth century. Indeed, two decades earlier it had received a significant stimulus by Lavoisier's discovery that all known biological materials contained at least carbon, hydrogen, and oxygen and by Berthollet's finding that most animal substances also contained nitrogen. Thereafter, proximate analyses of organic materials were increasingly supplemented by attempts to specify precisely the relevant proportions of these elements in a given sample. Lavoisier and Laplace also found that animal heat appeared to be completely accounted for by slow combustion of these substances, which fact suggested to many that not just respiration, but possibly all physiological processes, might well be purely chemical and physical in nature.[21]

---

18 *Autobiographical Notes*, pp. 38–39, 48–52, 170.   19 Ibid., pp. 59–62.
20 *Djurkemien* (see note 6); Söderbaum, *Levnadsteckning*, Vol. I, pp. 189–190, 198–200, 213–217, 229–231.
21 Excellent background to and discussion of these issues is provided by F. L. Holmes, "Elementary Analysis and the Origins of Physiological Chemistry," *Isis*, 1963, *54*:50–

Of course, biological materials were often extremely complex mixtures of extremely complex compounds, and although that complexity could be reduced by physical and chemical techniques, there was usually a limit to this simplification. Even when such techniques did result in the isolation of apparently pure homogeneous substances, unequivocal results were still elusive. Scheele's discovery of a large number of new organic acids seemed to spoil the elegance of Lavoisier's simple scheme of a mere handful of such substances, and for a generation after Scheele's work many chemists sought to show that many of these new acids were but impure mixtures or modifications of well-known acids.[22]

Furthermore, analysis of animal fluids proved discouragingly variable, as so many factors affect the exact composition of many of these materials. Xavier Bichat complained:

> [Inorganic substances], always the same, are known when they have been analyzed once with precision; but who could claim to know [animal fluids] after a single analysis, or even after several made under the same circumstances? Urine, saliva, bile, etc., taken randomly from this or that subject are analyzed, and from their examination animal chemistry arises; this is, if I may say so, [merely] the cadaveric anatomy of fluids.

In another work, he concluded: "Let us leave to chemistry its affinity, and to physics its elasticity [and] its gravity. Let us employ in physiology only sensibility and contractility."[23] In sum, when Berzelius began his career, animal chemistry was an identifiable discipline with a reasonably large number of empirical results, but the reliability of those results was often subject to doubt, and they were not undergirded by a secure theoretical or methodological base.

It might be taken as axiomatic that, however much philosophical and religious convictions guide scientific work in general, they must have an even greater influence on disciplines whose internal structure is still young and insecure. Such extrascientific concerns are even more to be expected in a discipline that impinges on the mysteries of life. Thus animal chemists, like physiologists, found it difficult to avoid these larger issues, nor

81. For similar detail with a focus on phytochemistry, see Reinhard Löw, *Pflanzenchemie zwischen Lavoisier und Liebig* (Munich: Donau-Verlag, 1977).

22 A. L. Lavoisier, *Elements of Chemistry*, tr. R. Kerr (Edinburgh, 1790), pp. 115–122. On this point see Evan M. Melhado, *Jacob Berzelius: The Emergence of His Chemical System* (Stockholm and Madison: Almqvist & Wiksell and University of Wisconsin Press, 1981), pp. 120–125, 139–141.

23 X. Bichat, *Recherches physiologiques sur la vie et la mort*, 3rd ed. (Paris, 1805), p. 76; Bichat, *Anatomie générale, appliqué à la physiologie et à la médecine*, 4 vols., nouvelle ed. (Paris, 1821), Vol. I, pp. 21–22.

indeed did they particularly desire to. The perennial debate between vitalists and reductionistic or mechanistic materialists is an example.

Eighteenth-century vitalist belief ranged from perhaps its strongest form in G. E. Stahl, who posited a transcendent agent directing the development of the organism, to that of Albrecht von Haller and Bichat, who suggested an immanent emergent vital force that was not independent of the organism and was essentially unknowable by human science. The other pole of thought may be represented by Julien de La Mettrie, who advocated a radical atheistic-materialistic world view.[24] These polarized positions were mediated by a number of eighteenth-century physiologists who creatively blended materialist and vitalist precepts. Théophile Bordeu (1722–1776), for instance, enunciated a strongly mechanical vision of the function of nerves and glands, denying the operation of a physiologically active "soul," while also affirming the necessity of an immanent – and material – vital action. This view was typical of the entire Montpellier school.[25] The convergence of vitalism and materialism can also be seen in the work of Pierre Jean Georges Cabanis at the end of the century.[26]

This convergence also gave rise to a tradition that pertains directly to our subject and has recently been traced by Timothy Lenoir. He calls the overall tradition teleomechanism, and its earliest variety vital materialism. This set of ideas was developed simultaneously and independently by Johann Friedrich Blumenbach (1752–1840) and by Johann Christian Reil (1759–1813), both of whom had been stimulated by Immanuel Kant's philosophy of biology. Blumenbach and Reil thought that Kant had been successful in showing how biological explanation could be both mechanistic and teleological without being mystical, namely, by suggesting the existence of a reflexive teleological (rather than the conventional linear-mechanical) mode of causation, which the human mind is incapable of comprehending in detail but which is also purely material. They posited an emergent *nisus formativus* (*Bildungstrieb* or formative force) that directs the development of the organism by this mode of causation and

24 For background on these issues, see J. Goodfield, *The Growth of Scientific Physiology* (London: Hutchinson, 1960); Nordenskiöld, *History of Biology;* Shirley Roe, *Matter, Life, and Generation: Eighteenth-Century Embryology and the Haller-Wolff Debate* (Cambridge: Cambridge University Press, 1981); and the relevant biographies in the *Dictionary of Scientific Biography*.

25 Elizabeth Haigh, "Vitalism, the Soul, and Sensibility: The Physiology of Théophile Bordeu," *Journal of the History of Medicine,* 1976, 31:30–41; Haigh, "The Roots of the Vitalism of Xavier Bichat," *Bulletin of the History of Medicine,* 1975, 49:72–86.

26 Martin S. Staum, *Cabanis: Enlightenment and Medical Philosophy in the French Revolution* (Princeton: Princeton University Press, 1980), pp. 30–37, 49, 78–93, 178–182, 201–206.

hence is both teleological and vitalistic. But both emphasized that what they referred to as a vital force must have a perfectly materialist basis, the proper complete physical picture of the organic realm remaining hidden only due to ignorance of its detailed operation.[27]

Regarding the possibility of artificial synthesis of organic material, there was a range of opinion at the beginning of the century.[28] No one had yet been able to accomplish such a synthesis, and most chemists agreed that it was either unlikely or impossible even in the future. In Fourcroy's enormously popular *Philosophie chimique* (1795) – the first work on chemistry that Berzelius studied – the author flatly stated the impossibility of organic synthesis. In the first edition of *A System of Chemistry* (1802), published just when Berzelius was graduating, Thomas Thomson argued the same point in detail, and posited a "superior Agent" or "power" that operates in the organism in a fashion that is neither physical, chemical, nor mechanical.[29]

As early as 1805, however, Charles Hatchett produced a substance from pit coal and nitric acid that appeared identical to tannin from oak galls. Although he regarded this work as a true synthesis of a vegetable from a mineral material, he did not emphasize this aspect, and thought the importance of the discovery to be economic rather than scientific. In a follow-up article he declared that the "synthesis of natural products . . . is . . . seldom to be obtained," but only because of the complexity of the mixtures on which one must operate.[30] About the same time E. J. Bouillon la Grange expressed confidence that he would soon be able to synthesize gallic acid.[31] And in 1808 Berzelius discovered that a material could be derived from the solution of cast iron in certain mineral acids that appeared identical to what was known as "extractive matter," an organic extract from decomposing vegetable matter. He noted the parallel to Hatchett's work, but, like Hatchett, expressed no particular surprise at the apparent "synthesis."[32]

27 Lenoir, *Strategy of Life*, especially pp. 9, 17–37.
28 The best discussions of nineteenth-century opinions on organic synthesis are Brooke, "Vital Force?" and Brooke, "Organic Synthesis."
29 A. F. Fourcroy, *Philosophie chimique*, 2nd ed. (Paris, An 3 [1795]), p. 125; T. Thomson, *A System of Chemistry*, 4 vols. (Edinburgh, 1802), Vol. IV, pp. 232, 463–464, 518–523. By 1830 he had changed his mind, as is noted by Brooke, "Vital Force?" pp. 88–89.
30 Charles Hatchett, "On an Artificial Substance Which Possesses the Principal Characteristic Properties of Tannin," *Philosophical Transactions of the Royal Society*, 1805, 95:211–224, 285–315; 1806, 96:109–146, especially pp. 224, 140.
31 E. J. Bouillon la Grange, "Faits pour servir à l'histoire de l'acide gallique," *Annales de chimie*, 1806, 60:156–184, on p. 184.
32 Berzelius, "Försök till tackjernets analys," *Afhandlingar i fysik, kemi och mineralogi*, 1810, 3:128–152; tr. as "An Attempt to Analyse Cast Iron," *Philosophical Magazine*,

The religious views of chemists were also relevant for their relation to the philosophy of biology. In the first third of the nineteenth century natural theology in particular was still extremely widespread and popular, representing both a mode of explanation and, for some, a compelling motivation for scientific work. This was especially the case for those who were exploring the wondrous intricacies and *Zweckmässigkeit* of physiology. Other than the association of atheism with reductionistic or mechanistic materialism, however, natural theology and other specific religious doctrines do not easily correlate with specific philosophies of biology.[33] Even atheists and freethinkers, such as Fourcroy, often used teleological explanations unapologetically, usually substituting "nature" for the natural theologian's "God."[34] The paucity of such correlations notwithstanding, religious commitments were still important for philosophical opinions (and vice versa): Quite simply, there was, and is, a variety of ways of making connections between these realms.

From his student days, such philosophical and religious concerns strongly influenced Berzelius's science. From his later references we may infer that he studied such authors as Haller in physiology, Bichat in anatomy, and Reil in medicine. It also appears that Berzelius was influenced, directly or indirectly, by the Montpellier vitalists. On graduation, Berzelius started in on research in animal chemistry with enormous confidence and energy. His first paper on this subject, published early in 1805 shortly after his and Hisinger's articles on electrolysis and cerium had appeared in foreign journals, concerned the red coloration of bones of animals that had been fed madder.[35] This paper was the harbinger of an awesome profusion of physiological-chemical studies flowing from Berzelius's laboratory. In the

1812, 40:245–258, on p. 248. Berzelius commented that these experiments were performed early in 1808. Other chemists in the first third of the century who countenanced at least the possibility of organic synthesis are mentioned by Brooke, "Vital Force?" pp. 88–89.

33 In *Matter, Life, and Generation,* Roe shows connections between the philosophies and religious ideas of Haller and C. F. Wolff. J. H. Brooke has described similar issues surrounding interpretations of natural theology, for example, in his "Natural Theology and the Plurality of Worlds," *Annals of Science,* 1977, 34:221–286. I am indebted to John Brooke for some perceptive comments on this subject in a private communication.

34 Nature's purpose in putrefaction, Fourcroy explained, is to resimplify complex organic substances in order to prepare them for reabsorption into young growing plants. He concluded (*Philosophie chimique,* pp. 162, 172–173) that nature's "perpetual circulation of compositions and decompositions . . . attest to her power and illustrate her fecundity, at the same time announcing a plan as grand as it is simple in its operations" (my translation). On Fourcroy, see William Smeaton, *Fourcroy: Chemist and Revolutionary, 1755–1809* (Cambridge: W. Heffer and Sons, for the author, 1962).

35 Berzelius, "Versuche über die Färbung der Thierknochen durch genossene Färberrothe," *Neues allgemeines Journal für Chemie,* 1805, 4:119–133, on pp. 119–121, 130–133.

following year, 1806, he published analyses of bone, fatty acids, and marrow in Swedish and a lengthy German paper on human feces. In the same year the first volume of his *Föreläsningar i djurkemien* appeared, the second volume following in 1808.

This remarkable work has never been translated, and is rarely discussed in the secondary literature,[36] but an understanding of its content is essential for following the evolution of Berzelius's ideas on animal chemistry. He wrote the work, he said, to provide a Swedish summary of his lectures for the benefit of advanced medical students, many of whom were weak in foreign languages – after all, they were attending a school of surgery, and not a university.[37]

The first volume was written at a time when Berzelius himself had done very little experimental research in this field; it is essentially a critical review of the discipline, with considerable philosophical content. The ideas of Haller, Bichat, and Reil are particularly thoroughly described and frequently criticized. He soon realized that animal chemistry could greatly benefit from his own efforts in the laboratory, and delayed the appearance of the second volume until he could complete an independent examination of a variety of topics that interested him.[38] The reporting of these results produced a very different, and considerably larger, second volume. He found, for instance, that lactic acid was indeed a distinct acid, as Scheele had claimed, and was not, as Fourcroy and other French chemists had thought to have demonstrated, a variety of acetic acid. He also was able to show that the coloring matter of blood was not, as Fourcroy thought, basic phosphate of iron dissolved in albumin, but was a compound of iron with a "peculiar animal substance" that was too complex to be fully characterized.[39]

As important as these factual results were, they must have made much less interesting reading for Berzelius's students than did the controversial first volume. The preface to this volume advertises the discipline as the "most interesting" field of chemistry and suggests that the present work was simply preparatory spadework for a future complete physiology that would eventually transform practical medicine. Previous writers on the subject had not understood this necessary *physiological* orientation, and so had wasted much of their time and energy.[40] Berzelius's

---

36 The only discussions in major languages are brief: Jorpes, *Life and Work*, pp. 34–37; Jørgensen, "Lebenskraft," pp. 265–268; and Nordenskiöld, *History*, pp. 371–374.

37 Berzelius, *Autobiographical Notes*, p. 61; *Djurkemien*, Vol. I, preface.

38 *Autobiographical Notes*, p. 62.

39 Fourcroy, *Système des connaissances chimiques*, 11 vols. (Paris: 1800–1802), Vol. IX, pp. 150–156; *Djurkemien*, Vol. II, pp. xi–xxxix, 430–441. These results were reported to J. G. Gahn early in 1808; *Bref*, Vol. IV:ii, pp. 23–25, 34–35.

40 Berzelius, *Djurkemien*, Vol. I, unpaginated preface. See also Jorpes, *Life and Work*, p. 7.

youthful hubris overlooked a similar concern in his contemporary colleagues in this field, such as Fourcroy, as well as his own mentor Pehr Afzelius – even if Berzelius did prove a better bench chemist than they.[41] His real opponents here, of course, were physiologists such as Bichat and Haller, who had no understanding of the utility of chemistry for their discipline.

The opening pages assert a fully materialist and unitary view of organic and inorganic substances. Organic and inorganic materials are composed of the same components and affinities and follow the same laws of nature; they can be transformed one into the other, and even spontaneous generation is not excluded. There is an "endless variation of natural products," but these are like different melodies using the same notes or languages using the same letters. All the forces that work together to create life must be nothing other than physical and chemical in nature.[42] An example of Berzelius's approach is his discussion of assimilation of food substances into animal tissues. The "affinity of aggregation" ("*sammanhopningsfrändskapen*"), which assimilates like materials into like tissues, is analogous to the seeding of a solution of mixed salts with a crystal of one of the components – only that one component precipitates out.[43] This conception reveals the probable influence of Reil, who described the process of assimilation in animals as "*thierische Krystallisation*," and who asserted numerous other analogies between organic and inorganic phenomena. It also echoes the even earlier "intussusception" theory of Louis Bourguet.[44]

According to Berzelius there are, however, limits to our ability to deduce the operation of those unitary laws in particular cases. For instance, the brain and nerves may well operate using electrical impulses, as had often been supposed, but there is as yet no proof of this, and there may

---

41 For example, Fourcroy and Vauquelin, "Memoire pour servir à l'histoire naturelle chimique et médicale de l'urine humaine . . . ," *Annales de chimie*, 1799, 32:80–162, on p. 150; Fourcroy, "On the Existence of Phosphate of Magnesia in Bones," *Philosophical Magazine*, 1806, 24:262–265, on p. 265. Fourcroy's concern for a physiological orientation for chemistry is discussed by Smeaton, *Fourcroy*; and by J. E. Lesch, *Science and Medicine in France: The Emergence of Experimental Physiology, 1790–1855* (Cambridge, MA: Harvard University Press, 1984), pp. 45–47.

42 Berzelius, *Djurkemien*, Vol. I, pp. 1–4.   43 Ibid., p. 176 and 176n.

44 J. C. Reil, "von der Lebenskraft," *Archiv für die Physiologie*, 1796, 1:8–162, on pp. 64–82; he cites other analogies between organic and inorganic nature on pp. 45–57. C. F. Wolff also attributed nutrition and vegetation to a chemical principle of like natures attracting each other, and secretion and excretion to the opposite; *Von der eigenthümlichen und wesentlichen Kraft der vegetabilischen sowohl als auch der animalischen Substanz* (St. Petersburg, 1789), pp. 39, 42, 53, 66–67; cited and discussed in Roe, *Matter, Life, and Generation*, pp. 115–118. I know of no evidence of a direct influence of Wolff on Reil or on Berzelius, but this possibility cannot be discounted. Bourguet's intussusception theory is discussed by Thomas Hankins, *Science and the Enlightenment* (Cambridge: Cambridge University Press, 1985), p. 128.

never be. Another possibly insoluble problem was the detailed mechanism of the action of the nerves on the organs that they stimulate to secrete or excrete; "chemistry cannot give us the least illumination on this matter."[45] But this limitation to our scientific powers did not in the least diminish Berzelius's faith in the materialist basis of physiology:

> It does not require much reflection to determine that all of the brain's functions are caused by and depend on all of the same physical and chemical laws as the other functions of the body. . . . As absurd and as paradoxical as this may seem . . . nonetheless, our judgment, our memory, our thoughts, indeed, all of the brain's functions, are fully as much organic-chemical processes as are those of the stomach, intestines, lungs, glands, etc.; but here chemistry rises to a higher realm, where our research shall never reach.[46]

As Nordenskiöld comments, even La Mettrie could not have expressed himself more clearly,[47] and one is reminded of the familiar image of Enlightenment physiology that the brain "secretes" thoughts like any other organ.[48] Karl Vogt's later scatological metaphor of the secretion of thoughts by the brain as analogous to the secretion of urine by the kidneys, of course, was much maligned, and even ridiculed.[49] Berzelius did not want to be misinterpreted. He attached a footnote to this passage:

> Some philosophers deny the independence of the soul, and view it as a product of organization, whose existence ceases with life. This manner of reasoning, which is so little in accordance with our hope and our essentially inborn feeling of the immortality of the soul, is called materialism.[50]

Clearly, Berzelius wanted to place some distance between himself and La Mettrie's form of atheistic materialism, but this footnote is by no means a ringing denunciation of materialism in general. These passages suggest that Berzelius was searching for a modus vivendi between radical

45 Berzelius, *Djurkemien*, Vol. I, pp. 91–92. Berzelius did, however, demonstrate (*Philosophical Magazine*, 1810, 35:108n) that severing an organ's nerve led to cessation of all secretions. This demonstration was by no means new; see Haigh, "Bordeu," pp. 34–35.

46 *Djurkemien*, Vol. I, p. 78; (my translation).   47 Nordenskiöld, *History*, p. 373.

48 See Staum, *Cabanis*, pp. 7, 92, 201–206. Cabanis's statement of this idea in his *Rapports du physique et du moral de l'homme* (Paris, 1802), a likely source for Berzelius, received considerable posthumous attention – and notoriety.

49 K. Vogt, *Physiologische Briefe* (Stuttgart and Tübingen, 1845–1847), p. 206, cited by F. Gregory, *Scientific Materialism in Nineteenth-Century Germany* (Dordrecht: Reidel, 1977), p. 64.

50 Berzelius, *Djurkemien*, Vol. I, p. 78n; (my translation).

and unpalatable alternatives, a search along a similar path that the Montpellier vitalists and materialists of such varied opinions as Charles Bonnet and Pierre Cabanis had traveled – a search that by no means sought to exclude the existence of *some* sort of vital function and of a human soul.

More particularly, there is reason to believe that Berzelius had become convinced by the principal of doctrines of vital materialism. One of the chief architects of this tradition, Reil, is frequently mentioned in the first volume of Berzelius's animal chemistry – though to be sure Berzelius by no means always found himself in agreement with Reil's opinions – and Reil's *Archiv für Physiologie* is cited in the preface as indispensable, along with two other journals. The first volume of Reil's *Archiv* (1796) contained a long essay on vital forces where doctrines of vital materialism are clearly espoused.[51] No obvious connections between Berzelius and Kant or Blumenbach have been found, but one of Blumenbach's students was Girtanner, whose writings provided Berzelius's introduction to antiphlogistic chemistry. Girtanner had also written a vital-materialist treatise, *Ueber das Kantische Prinzip für die Naturgeschichte*. Essentially a brief for the vital materialism of Blumenbach, it was dedicated to his teacher and was published in the same year as Reil's essay.[52] It would be surprising if Berzelius had not read it.

There are indeed elements of vital materialism in Berzelius's animal chemistry. Its materialist tendencies have already been noted, as well as the conviction of a certain opacity to human research into the detailed workings of nature. However, if one were to securely classify Berzelius as a vital materialist one might expect to see more of the language of teleology and of vitalism and at least a hint of the Kantian reflexive mode of causation.

These additional elements are perceptible in two summaries of the field of animal chemistry that Berzelius prepared soon after his lectures were published. The first of these was a detailed overview of his own work on the composition of animal fluids, in preparation since 1808 and published in his and Hisinger's *Afhandlingar i fysik* for 1810.[53] He wrote to Hisinger early in 1810: "Don't get tired of my long snot-bile-urine paper, for it has truly great interest, even if the number of readers cannot be

---

51 Reil, "Lebenskraft" (published in July 1795).
52 C. Girtanner, *Ueber das Kantische Prinzip für die Naturgeschichte* (Göttingen, 1796). For a discussion of Reil, Girtanner, and vital materialism, see Lenoir, *Strategy of Life*, pp. 34–37; and Lenoir, "Kant, Blumenbach, and Vital Materialism," *Isis*, 1980, 71:77–108, especially pp. 96–99.
53 Berzelius, "Försök till en allmän öfverblick af varmblodiga djurvätskors sammansättning . . . ," *Afhandlingar i fysik, kemi och mineralogie*, 1810, 3:1–105, translated as "General Views of the Composition of Animal Fluids," *Medico-Chirurgical Transactions*, 1812, 3:198–276, on p. 256.

large."[54] Although much of this paper consists simply of analytical results, there are a number of passages of more general significance. In discussing the composition of milk, for example, Berzelius commented:

> Cheese, which is destined to be part of the nourishment of the young animal, has very peculiar characters, which, as it would seem, fit it for this office [, especially the property of dissolving calcium phosphate]. We may conclude that nature has thus sought to assist the digestive powers of the young during a period of their lives, in which there exists in the oeconomy the greatest demand for earthy phosphates for the purposes of ossification, which is at that time advancing so rapidly.[55]

The presence of lactic acid in the urine was also significant for Berzelius, as, "If I may be allowed to speculate on final causes, I should say that it is destined to hold the earthy phosphates in solution, and obviate the dire effects of their deposition in a solid mass."[56] A similar passage treats of "the organized laboratory which nature employs" to convert blood, a fluid that forms the common starting material for all body fluids, into secretions unique to each organ.

> Not only is the chemical agent which produces these changes unknown to us, but we shall in vain search for any analogous chemical operation. It is doubtless easy to conjecture, that it is by the influence of the nervous system, that this decomposition of blood into the secreted fluids is effected; but what is this influence? If electric, how can it be brought to accord with our present knowledge of electric agency? But [we should avoid] vain conjectures on a subject which perhaps will ever remain a mystery.[57]

Taken out of context, these passages may appear to exemplify the teleology of traditional natural theology, or even of a freethinker like Fourcroy; but the prominent materialist tendency suggests a direct or indirect influence of Kantian vital materialism.

In August 1810, Berzelius delivered an address, according to custom, on relinquishing the presidency of the Royal Swedish Academy of Sciences. He chose to summarize the "Progress and Present State of Animal Chemistry."[58] The opening of this address reiterated a common theme of his earlier writings on this subject: The body is a laboratory that uses chemical processes to produce the phenomena of life, even if the ultimate

54 *Bref,* Vol. IV:i, p. 30.   55 Berzelius, "Animal Fluids," pp. 274–275.
56 Ibid., p. 262.   57 Ibid., p. 234.
58 Berzelius, "*Öfversigt af djur-kemiens framsteg och närvarande tillstånd* (Stockholm, 1812); *A View of the Progress and Present State of Animal Chemistry,* tr. G. Brunnmark (London, 1813).

cause and detailed operation of those phenomena are and will remain inscrutable.

> We call this hidden cause *vital power*; and like many others, who before us have in vain directed their deluded attention to this point, we make use of a *word* to which we can affix no idea. This *power to live* belongs not to the constituent parts of our bodies, nor does it belong to them as an instrument, neither is it a simple power; but [is] the result of the mutual operation of the instruments and rudiments on one another – a result, which varies as the operations vary, and which often, from small changes and obstructions, ceases altogether.[59]

This passage clearly expresses Berzelius's position regarding the relationship between animate and inanimate phenomena. True to his unitary standpoint, he continued to affirm that organic and inorganic nature are composed of the same components and follow the same physical laws; but he also affirmed that there was nevertheless something unique about living creatures and suggested that the peculiarity of life inhered in the reflexive teleological mode of causation that was being urged by the vital materialists. Berzelius pronounced this same view 21 years later, when he expressed disagreement with C. A. Argadh's vitalism:

> To assume that the elements of organic nature are endowed with fundamental forces other than [those] in the inorganic realm is an absurdity. The difference in the products rests on the different circumstances under which the forces work, and these can vary endlessly; but [the fact] that we cannot correctly conceive the circumstances which occur in organic nature gives us no sufficient reason to hypothesize other forces.

Berzelius represented the same position publicly a few years later (1838) in a discussion of "some questions of the day in organic chemistry."[60]

In 1810 (and afterward) Berzelius insisted on the present inaccessibility of these "circumstances" of organic nature; he used the operation of the brain and nervous system as the best exemplar of this point of view. Again adducing the inexplicable fact that various organs under the influence of the nerves produce various secretions from the same blood, he averred that until we are able to produce analogous phenomena in vitro, "we shall never be able to discover the laws of those operations." It is

---

59 *Animal Chemistry*, pp. 4–5.
60 Berzelius to C. A. Agardh, 22 November 1831, *Bref*, Vol. IV:iii, pp. 71–72; Berzelius, "Om några af dagens frågor i den organiska kemien," *KVA, Handlingar*, 1838:77–111, on p. 78.

indeed astounding that our thoughts are produced by chemical reactions in the brain, and yet this is an "incontrovertible truth." "But is it not probable, that human understanding . . . which has . . . attained a degree of perfection, the summit of which is concentrated in GOD, may one day explore itself and its nature? I am convinced it will not."[61]

Berzelius's position, and that of the other vital materialists, represented an ingenious compromise between two unpalatable alternatives. Thoroughgoing rationalism could lead to the slippery slope of atheistic materialism – as it led to the scientific materialism of the later nineteenth century – and it also seemed to ignore the undeniable and conspicuous differences between the two realms of nature. Conversely, too strong a vitalist philosophy had elements of mysticism, theism, and unfounded hypothesis that were anathema to representatives of the late Enlightenment. Vital materialism provided a modus vivendi between these two alternatives.

In 1806 Berzelius decided that an advanced text on physiological chemistry did not suffice for his teaching needs, as many of his students needed a much more elementary grounding in the principles of general chemistry. He therefore began a task that would occupy him off and on for the rest of his life, his *Lärbok i kemien* [Textbook of chemistry]. It would prove to be a momentous event for the future direction of his research. He commented later that "animal chemistry would certainly have been my principal interest permanently" had he not been compelled to write this textbook.[62] His preparatory literature review turned up the work of J. B. Richter, which led to a new interest in stoichiometry; by June 1809 Berzelius had discovered Daltonian chemical atomism, and he began to move aggressively in this new direction. In 1810 he published a landmark study in Swedish that definitively established the laws of equivalent and multiple proportions as general and precise laws of inorganic nature. He then began to work on a series of follow-up projects.[63]

At this time Berzelius's foreign reputation was by no means commensurate with the immense volume and quality of his research, mostly because the large majority of it was still available only in Swedish. The circumstance began to change in consequence of his first trip abroad, to

61 Berzelius, *Animal Chemistry*, pp. 5–8, on p. 8.
62 *Autobiographical Notes*, pp. 62–64, 171, on p. 171; *Bref*, Vol. IV:i, pp. 19–20; Jorpes, *Life and Work*, pp. 37–42.
63 The 1810 study appeared in German as "Versuch, die bestimmten und einfachen Verhältnisse aufzufinden, nach welchen die Bestandtheile der unorganischen Natur mit einander verbunden sind," *Annalen der Physik*, 1811, 37:249–334, 415–472; 38:161–226; 1812, 40:162–208, 235–320. For these events, see Anders Lundgren, *Berzelius och den kemiska atomteorin* (Uppsala: Almqvist & Wiksell, 1979), and A. J. Rocke, *Chemical Atomism in the Nineteenth Century* (Columbus: Ohio State University Press, 1984), pp. 66–78.

England in the summer of 1812. The trip was a fortunate one in every respect. England was the only country where chemical atomism was then being pursued (and not just by John Dalton, but also by Thomas Thomson, Humphry Davy, and William Wollaston), and it also had a healthy and active community of animal chemists.[64] Since 1808 Berzelius had been corresponding with Davy about a variety of scientific subjects. He sent Davy a copy of his *Föreläsningar i djurkemien,* and Davy immediately began to arrange for its translation into English. This translation was nearly finished but never published; however, it did circulate in interested groups in London.[65] In fact, it precipitated hurried publication of research by Wollaston and Everard Home on the electrical character of nerve impulses and their relation to secretions. It turned out that Thomas Young was also pursuing very similar research.[66]

Berzelius was befriended by Alexandre Marcet, an expatriate Swiss chemist who was also active in animal chemistry, and at Marcet's urging English translations of Berzelius's two summaries of 1810 were quickly prepared. One of these, the summary of his own experimental work in the field, was completed in time for Marcet to read it at a meeting of the London Medico-Chirurgical Society, with Berzelius in the audience.[67] The reason for haste was made clear when John Bostock read a remarkably similar paper to the same society six months later. Bostock noted that his experiments dated from as early as 1805, and that the paper was mostly written by the time he saw Berzelius's work.[68] Berzelius's second summary, the speech to the Royal Swedish Academy of Sciences, appeared as an English offprint in 1813. In both cases these were the first detailed treatments of Berzelius's discoveries and opinions in the field to appear in a major language, and the English versions were soon retranslated and published in German and French journals.[69] They were received

64 N. G. Coley, "The Animal Chemistry Club: Assistant Society to the Royal Society," *Notes and Records of the Royal Society of London,* 1967, 22:173–185, especially pp. 177–179.

65 See letters in *Bref,* Vol. I:ii, pp. 9–10, 17–18. There were equally abortive attempts to produce French and German translations between 1810 and 1812 (ibid., Vol. I:i, pp. 5, 11–12, 14–15, 25, 36). A second attempt between 1812 and 1814 to have an English translation prepared, this time with Marcet playing the leading role in London, likewise came to naught (ibid., Vol. I:iii, pp. 13, 24, 45–46, 49, 64, 91–92, 97–98).

66 W. H. Wollaston, "On the Agency of Electricity on Animal Secretions," *Philosophical Magazine,* 1809, 33:488–490; E. Home, "Hints on the Subject of Animal Secretions," *Philosophical Transactions of the Royal Society,* 1809, 99:385–391, on p. 385n.

67 E. Wöhler, tr., "Aust Berzelius' Tagebuch während seines Aufenthaltes in London im Sommer 1812," *Zeitschrift für angewandte Chemie,* 1905, 18:1946–1948; 1906, 19:187–190, 571–576; on pp. 573, 575.

68 J. Bostock, "On the Nature and Analysis of Animal Fluids," *Medico-Chirurgical Transactions,* 1813, 4:53–88.

69 See Arne Holmberg, *Bibliografi över J. J. Berzelius* (Stockholm: Almqvist & Wiksell,

extremely favorably, especially in England.[70] Berzelius thus belatedly acquired a justified reputation as the leading European physiological chemist.

This belated recognition is somewhat ironic considering the fact that by this time Berzelius had turned away from animal chemistry per se[71] and was pursuing stoichiometry and atomic theory with his full attention. This switch of specialty, however, might be viewed more as a *reculer pour mieux sauter* than as a definitive change of direction, for there is evidence that one of the sources of Berzelius's interest in this new subject was for what it could ultimately offer physiological chemistry. For instance, he thought that iron was present in the coloring matter of blood in the form of a lower oxide that was as yet unknown in inorganic chemistry; he used this supposition as a model for seeking – and simply hypothesizing when the search proved fruitless – lower oxides of other metals in inorganic compounds. In this way research in inorganic chemistry could, potentially at least, be used to clarify organic phenomena, and vice versa.[72]

But this strategy was applied in a much broader fashion as well. Berzelius had learned to use inorganic stoichiometry to define and identify pure substances and to subsume them into comprehensive theories of chemical composition. Such an enormously useful tool ought to be able to be turned to organic nature as well. A recent scholar has suggested that one of Berzelius's signal innovations was to stress that organic compounds must be identified in terms of specific well-defined substances, as opposed to generic mixtures; he argues further that one of Berzelius's first "deployments" of stoichiometry had just this intent.[73] Indeed, we find that one of Berzelius's earliest attempts at quantitative elemental analysis subsequent to his discovery of stoichiometry was perhaps the most basic organic compound of all, acetic acid. The attempt was not a

1933), pp. 20–27. In addition to the English and Swedish versions, five German and five French translations and reprints of these two works appeared between 1813 and 1815.

70 See letters in *Bref*, Vol. I:iii, pp. 19–20, 45–46, 58, 64, 120–121; Vol. IV:ii, p. 77.

71 Jorpes writes (*Life and Work*, p. 37) that "Berzelius soon became disgusted with the general unreliability of the analysis of animal products, and said in 1810 that he had written his last paper on nasal mucus, bile, urine, and related topics." Though I have not been able to supply a precise reference to Jorpes's comment, it does seem consistent with Berzelius's writing two capstone summaries in that year and then devoting his energies to a somewhat different (and certainly more reliable and precise) topic, stoichiometry and chemical atomism. It would seem, then, that he finally capitulated to at least one of Bichat's opinions. But see below here for further details.

72 Berzelius, "Versuch," 37:323, 465–472. This matter is discussed and perceptively analyzed by Melhado, *Berzelius*, pp. 173–191

73 Melhado, *Berzelius*, pp. 54–57, Ch. 6, et passim.

success and was never published. His attempts in 1808 to electrolyze organic salts after analogy with inorganic electrolyses were also unsuccessful, but once more reveal his conviction of unity between the organic and inorganic realms.[74]

About September 1810, Berzelius discovered, as a product of his stoichiometric work, an empirical generalization that was to guide his research for many years in the future – the oxygen content of acid component of any inorganic salt is a small integral multiple of that of the base.[75] He soon concluded, with limited experimental evidence to support such a view, that organic acids must also follow this "oxide rule." He then enunciated his "principle of formation of organic compounds": Organic compounds consist of at least two combustible elements combined in common with a portion of oxygen that suffices to oxidize fully only one of the components, and this compound cannot be separated into proximate components without destruction of the whole. Whenever we find such a substance, we can confidently ascribe it to organic origins – although Berzelius readily conceded that there were two known examples of synthetic materials exhibiting these characteristics, Hatchett's artificial tannin and his own artificial extractive matter.[76] In these beliefs Berzelius was echoing ideas of his great predecessor Lavoisier.[77]

Berzelius laid great stress on the importance of his oxide rule for organic chemistry, as, in modern language, it could often be used to calculate molecular formulas from empirical formulas. This sort of calculation was particularly critical for organic molecules, where the average numbers of atoms were much larger than in inorganic formulas, making the conversion of percentage composition into empirical formulas and then into molecular formulas all the trickier. He tried to follow up on this perceived breakthrough with another attempt to analyze simple organic acids, but once again he had little success. He wrote Gay-Lussac: "A hundred times I have begun experiments which I have always then abandoned, despairing of being able to come to exact results; and as soon as I attempted to combine calculations with experiment, my hopes were further diminished." Part of the problem, he recognized, was with his

74 These early attempts are mentioned in Berzelius, "Versuch," 37:471–472; and Berzelius, "Proportions of Organic Nature," 4:403.
75 "Schreiben des Herrn Prof. Berzelius . . . über ein zweites neues Gesetz . . . 1. October 1810," *Annalen der Physik* 1811, 37:208–220; for a detailed discussion, see Melhado, *Berzelius*, pp. 179–184.
76 Berzelius, "Versuch," 38:217, 224–225; Berzelius, "Essai sur la nomenclature chimique," *Journal de physique*, 1811, 73:253–286, on pp. 260–261; discussed by Melhado, *Berzelius*, pp. 282–293.
77 See Melhado, *Berzelius*, pp. 122–141; and F. L. Holmes, *Lavoisier and the Chemistry of Life* (Madison: University of Wisconsin Press, 1985), pp. 261–409.

method of oxidation, which was too rapid, and his oxidizing agent, the superoxide (dioxide) of lead, which was difficult to purify. He also found it difficult to purify and dry his experimental samples sufficiently.[78]

Since 1805 Berzelius repeatedly registered his faith in the unity of nature and his consequent conviction that the laws of the inorganic realm would be found to apply to the organic as well.[79] The emergence of stoichiometry and chemical atomism marked an important watershed in Berzelius's career. His first research program after this discovery was to verify and elaborate these new conceptions in inorganic chemistry, but he was also impatient to apply the ideas to organic nature as soon as possible. With the acquisition of the oxide rule to assist the rest of the chemical atomist program, Berzelius now felt he had found the key that would unlock at least some of those closely guarded secrets of nature.

Unfortunately, and ironically, it was just at this time (1811–1813) that Berzelius discovered certain characteristics of organic compounds that threatened to destroy the most essential analogies between the two realms. One problem was that organic molecules seemed always to contain more than one atom of every constituent element. Berzelius had thought that he had discovered that in inorganic compounds one element is always represented as a single atom (a claim that I have elsewhere termed the single-atom axiom).[80] He thought that denial of the single-atom axiom would destroy chemical atomism, as it would open the door to an infinite number of possible formulas, with no way to decide between them. He encountered a second problem when he thought he found a very small but distinct amount of hydrogen in (anhydrous) oxalic acid, which would lead to a molecular formula with an unacceptably large number of carbon and oxygen atoms: How could one envision a molecule such as $C_{27}O_{18}H$?[81] A third problem arose when he began to develop his ideas on electrochemical dualism, as organic compounds simply did not seem to follow the same electrochemical regularities as inorganic substances.[82]

Still, his unitary faith was not shaken. In the preface to the second volume of his *Lärbok,* dated 1 March 1812, Berzelius emphasized the unique character of organic chemistry and the large number of differ-

78 Berzelius, "Versuch," 40:246–253; "Proportions of Organic Nature," 4:403; Berzelius to Gay-Lussac, 25 September 1811, *Bref,* Vol. III:ii, p. 116.
79 Berzelius's use of analogies from inorganic to organic chemistry is a central theme of Brooke, "Organic Synthesis"; see also John Brooke in this volume, Chapter 8.
80 Rocke, *Chemical Atomism,* p. 76. Melhado, *Berzelius,* p. 270, refers to this as the "law of the unitary atom."
81 Berzelius, "Cause of Proportions," pp. 447, 449–451; Berzelius, "An Address to Those Chemists Who Wish to Examine the Laws of Chemical Proportions . . . ," ibid., 1815, 5:122–131, on pp. 124 and 128. For discussions, see Rocke, *Chemical Atomism,* pp. 76, 95n., 158–163; and Melhado, *Berzelius,* pp. 270–278.
82 Berzelius, "Nomenclature chimique," pp. 260–261; Melhado, *Berzelius,* pp. 286–293.

ences from the mineral realm. However, he did not doubt that chemical atomism and electrochemistry would "gradually throw ever more light even over organic chemistry." The field at that moment was in a "critical period of transition," and so it would be necessary to delay the appearance of later volumes of the textbook that would contain the material on organic chemistry until more precise investigations had yielded fruit.[83]

The industrious Swedish chemist did not delay long before beginning those more precise investigations. On 12 October 1813 Berzelius reported to Marcet that he was that day beginning his fall and winter research, and the subject was a third and hopefully ultimate attack on organic elemental analysis.[84] He had some reason to be optimistic because Gay-Lussac and Thenard had recently published some organic analyses of their own, using a different oxidizing agent, potassium chlorate.[85] Berzelius borrowed this improvement and made some additional ones of his own. This was a subject close to his heart, and he worked with characteristic energy, tenacity, and brilliance through the cold Swedish winter months.[86]

Still, progress came with discouraging slowness. Early in December he wrote to J. G. Gahn that two months of heartbreaking labor had gone for naught. In despair, he often considered giving up the attempt completely. The first sign of success came in a subsequent letter to Marcet dated 7 January 1814, where he reported definitive analyses of seven organic acids. Six days later he told Gahn that he "nearly believe[d]" he had discovered the laws of chemical proportions for organic compounds, and that the oxide rule had been absolutely indispensable.[87] In April, transported with joy at his success, he wrote to Marcet that this was

> the most difficult project I have ever worked on, but also I hope that it is the most important of all that I have hitherto done and perhaps even of all that I will ever do. The study of organic nature sheds light on inorganic nature to a completely unexpected degree. You will see how chemistry will become clear and how its theoretical part will be consolidated by this study.[88]

---

83 Berzelius, *Lärbok i kemien*, Vol. II (Stockholm, 1812), pp. ii–iii.
84 *Bref*, Vol. I:iii, p. 77.
85 J. L. Gay-Lussac and L. J. Thenard, *Recherches physico-chimiques*, 2 vols. (Paris, 1811), Vol. II, pp. 265–350.
86 Berzelius described his modifications of Gay-Lussac's and Thenard's method in "Proportions of Organic Nature," 4:401–408. He also mentioned in another context (5:182) that most of his analyses were performed at a temperature of from zero to three degrees Celsius!
87 *Bref*, Vol. IV:ii, pp. 93–96, on p. 95; Vol. I:iii, pp. 88–89.
88 Ibid., Vol. I:iii, p. 94.

Berzelius's paper was published serially in Thomson's *Annals of Philosophy,* beginning in November of that year.[89] Berzelius provided not only highly accurate elemental analyses of eight organic acids, two sugars, potato starch, gum arabic, and tannin, but he also calculated molecular formulas for each compound, verifying the applicability for organic chemistry of chemical atomism and the oxide rule. Emboldened by this breakthrough in proving the essential analogies between the two realms of nature, he no longer found the undoubted dissimilarities so troublesome. Berzelius's "principle of organic formation," enunciated three years earlier with little empirical justification, emphasized some of those differences and was now fully established. To be sure, virtually never does a combustible element enter into an organic molecule in a single atom, but this difference from inorganic chemistry should not be surprising considering the manifold differences between the two realms; indeed, the non-applicability of the single-atom axiom allows organic nature to achieve its wondrous diversity. The apparent anomaly of oxalic acid, Berzelius noted, had been based partly on an erroneous analysis; besides, we should not be surprised to see such differences between organic and inorganic chemistry.[90]

There were some significant new ideas in this paper. Berzelius averred that "in all probability art will never be able to imitate" organic compounds, since

> at the instant of the formation of each ternary and quaternary oxide in organic nature, its elements receive in combining a new electro-chemical modification, on which their chemical properties chiefly depend. . . . It appears clear that in animals nature employs the nervous system to produce these new electro-chemical modifications, and to determine the nature and composition of the various substances produced in different parts of the animal body.[91]

Organic compounds are of course also formed in animals that appear to be devoid of a nervous system and in plants; but perhaps these organisms have internal anatomical structures analogous to true nervous systems, and only future research will be able to decide this point.

This major paper formed another important transition in Berzelius's career, for immediately after its publication he turned to mineral analysis and discovered an aptitude and enjoyment in this field that had always hitherto been lacking.[92] The motivation for this shift will not be explored here, but it is relevant to note that he had achieved an important mile-

89 Berzelius, "Proportions of Organic Nature."   90 Ibid., 4:323–329; 5:98 and 98n.
91 Ibid., 4:323, 329.
92 Berzelius, *Autobiographical Notes,* pp. 89–93, 178; *Bref,* Vol. IV:ii, pp. 97–99; discussed in Melhado, *Berzelius,* p. 264n.

stone in his attack on the chemistry of living organisms. Starting from the scientifically hazardous analysis of complex biological materials, he had succeeded in showing the way to the specific level of well-defined and homogeneous organic compounds,[93] and he had integrated the study of these substances into the revolutionary new theoretical structure of the atomic theory. In doing all of this he provided a new demonstration of their distinctness from inorganic compounds while simultaneously proving that the same general laws of chemical proportions applied to both realms of nature.

Since his days as an undergraduate student at Uppsala, Berzelius charted a scientific career that would apply a state-of-the-art understanding of chemistry to anatomy, medicine, and especially physiology. His route to this goal was through animal chemistry, a discipline that already had been and was being vigorously pursued by a number of capable scientists, but to which Berzelius himself was soon able to make outstanding contributions. While working in this field he encountered stoichiometry and atomic theory, which provided a new vision of chemistry in general and was to transform the chemical study of organic nature in particular.

By 1814 Berzelius was able to mark a milestone in pursuit of this new vision. His paper published in that year exhibits a number of characteristics that appeared for the first time in a single investigation, especially the precise elemental analyses of a number of organic substances, followed by calculation of what we would now call both empirical and molecular formulas, using procedures for the determination of atomic weights and other chemical lemmas to assist in those calculations. There are even speculations on molecular structures and reaction mechanisms in this paper,[94] though it is beyond the scope of this chapter to look at these additional new ideas in detail. In short, in the space of the decade of chemical labor that we have examined we can discern the transformation of a research program in animal chemistry into a recognizably modern example of organic chemistry.

After 1814 Berzelius's career went in somewhat different directions, but the pattern he had done so much to establish persisted and was further developed by others as well. In particular, important elements of vital materialism are perceptible throughout the remainder of Berzelius's career, as are also some strong hints of natural-theological teleology. In one paper he came very close to an orthodox vitalist position: A "far more inscrutable foreign Power," a "primum movens which lies outside of the elements" was required to initiate life on earth.[95]

93 Melhado, *Berzelius*, pp. 54–57, Ch. 6, et passim.
94 Berzelius, "Proportions of Organic Nature," 5:98 and 98n, 262, 274–275.
95 Berzelius, "Versuch eines rein chemischen Mineralsystems," *Journal für Chemie und*

Similar ideas are found in his first systematic treatment of the entire field of organic chemistry, published in 1827. It begins with a long philosophical and theological discussion of the differences between the organic and inorganic realms. "An organism considered as an object of chemical research," he wrote, "is a factory in which many chemical processes happen, whose end result is to produce all those phenomena whose totality we call life." He emphasized here the same fixation that remained a constant throughout his career: The same laws govern both organic and inorganic nature. There are, however, undeniable differences between the two realms. This uniqueness of an organic substance

> does not consist in its inorganic elements, but rather in something different, which disposes the inorganic elements common to all living creatures toward the production of a certain result, characteristic and unique to each species. This *something*, which we call vital force, lies entirely outside the inorganic elements, and is not one of their original properties, such as gravity, imponderability, electrical polarity, etc.; but what it is, how it arises and ends, we do not understand. . . . A force which is both foreign to the inorganic world and incomprehensible to us introduced this *something* at some point; and not in such a way as if it were the result of chance, but rather in an admirable multiplicity, and designed with the greatest wisdom for certain purposes, for a ceaseless succession of mortal individuals which arise from one another.[96]

In case anyone missed the point, he emphasized once more his disdain for atheistic materialism:

> More than once, however, it was a part of the imagined depth of a shortsighted philosophy to assume everything as the work of chance. . . . But this philosophy did not perceive that what it assumed under the name of chance in dead matter is a physical impossibility. . . . [O]ur research discovers increasingly every day the structure of organic bodies arranged in such an astonishing way for certain final purposes, and it will always do us more honor to admire the wisdom which we cannot follow, than to philosophize in our pedantic arrogance regarding a supposed knowledge which perhaps is not given to us ever to understand.[97]

Physik, 1815, *15*:301–363, 419–451, on pp. 309–310. Cited (in German) in Melhado, *Berzelius*, p. 323.

96 Berzelius, *Lehrbuch der Chemie*, Vol. III:i 2nd ed. (Dresden, 1827), pp. 135–138; 5 ed. (Leipzig, 1856), Vol. IV, pp. 1–4; [my translations].

97 Ibid.

In all subsequent revised German and French editions of his textbook he repeated this introduction verbatim. However, he appears to have become concerned about the excessively theological tone of this passage, for in the last German and French editions he added some clarifying comments. Vital force, he wrote, is usually regarded as a unique force that suspends and alters the normal inorganic forces; but this view is false. Organic and inorganic nature follow equally the identical and inalienable fundamental forces of matter; what is unique to organic substances is the arrangement of "circumstances" accomplished by living organs, and, occasionally, by human art. This unknown something, which we call vital force (*innewohnende Kraft, force innée, nisus formativus*) in order to disguise our ignorance, thus represents the mysterious arrangement of uncounted different circumstances to produce the phenomenon of life, but uses only familiar inorganic forces.[98]

Throughout his life Berzelius exhibited a creative ambivalence toward these fundamental philosophical issues. It was probably the theistic elements in his personal background that led him to an unqualified rejection of atheistic or reductionistic materialism and the wholehearted acceptance of a limit to the ability of the human mind to penetrate the ultimate secrets of God's creation. And yet he was no less committed to a unitary, empirical, and rationalistic world view, a predilection that led him to scorn excessive speculation, *Naturphilosophie,* or what seemed to him to be a naïve and self-satisfied form of vitalism.

98 Ibid., 5 ed., Vol. IV, pp. 5–6; *Traité de chimie,* 2nd French ed. (Paris, 1849), Vol. IV, pp. 5–6. Berzelius's first complete treatment of organic chemistry appeared in the second German edition of his textbook (8 vols. in 4, Dresden, 1825–1831); this underwent two complete revisions, once for the third edition (10 vols., Dresden and Leipzig, 1833–1841) and again for the fifth (5 vols., incomplete, Dresden and Leipzig, 1843–1848). The 1856 edition (5 vols., Leipzig) is an unchanged, less expensive edition. There are only minor changes from the German editions for the two principal French editions of 1829–1833 (8 vols., Paris) and 1845–1950, 2nd ed. (8 vols., Paris, incomplete).

6

Novelty and Tradition in the Chemistry of Berzelius (1803–1819)

EVAN M. MELHADO

The image of Berzelius and his chemistry that has prevailed until recently issued from a nineteenth-century historiographical tradition. Its authors were mindful of the towering position that Berzelius enjoyed in his day, but paradoxically they cast doubt on his originality. This perspective characterizes the works of Meyer and Kopp (two of the leading accounts of nineteenth-century chemistry), it is shared by Söderbaum's book (appearing at the turn of the century), and it colored much more recent treatments, including those of Partington, Prandtl, and Jorpes.[1] Typically, the Berzelius portrayed in this literature corresponded to the perceptions held of their science by the chemists-cum-historian who chronicled its development. However, a new assessment of Berzelius's achievements and originality in chemistry has been emerging from more recent investigations by historians of science.[2] This study begins by briefly analyzing the traditional picture of Berzelius; it then suggests, in the light of recent scholarship, that the goal of Berzelius's system, discriminating species of matter, can be traced to several chemical traditions of the eighteenth century; it identifies the novelties that Berzelius introduced in the first half of his career as he pursued this goal (targeting for this purpose

1 Ernst von Meyer, *Geschichte der Chemie von den ältesten Zeit bis zur Gegenwart* (Leipzig, 1899), pp. 165–230; Hermann Kopp, *Geschichte der Chemie in der neueren Zeit* (Munich, 1873), pp. 311–333, 418–432, 498–517, 544–581, 600–631; H. G. Söderbaum, *Berzelius' Werden und Wachsen* (Leipzig, 1899); James R. Partington, *A History of Chemistry*, 4 vols. (1961–1970), Vol. IV, Ch. 5; Wilhelm Prandtl, *Humphry Davy, Jöns Jacob Berzelius: Zwei führende Chemiker aus der ersten Hälfte des 19. Jahrhunderts* (Stuttgart, 1948); J. Erik Jorpes, *Jac. Berzelius: His Life and Work* (Berkeley, 1970).

2 This chapter draws particularly on Evan M. Melhado, *Jacob Berzelius: The Emergence and Development of His Chemical System* (Madison and Stockholm, 1981); Melhado, "Mitscherlich's Discovery of Isomorphism," *Historical Studies in the Physical Sciences*, 1980, 11:87–123; Melhado, "Mineralogy and the Autonomy of Chemistry around 1800," *Lychnos* 1990:229–262; and Anders Lundgren, *Berzelius och den kemiska atomteorin* (Uppsala, 1979).

Berzelius's *Proportions chimiques* of 1819),[3] and it concludes by providing an enhanced image of Berzelius's originality.

The Nineteenth-Century Image

In developing their view of Berzelius, traditional authors, who belonged to the generation or two following Berzelius, were mindful of the stark contrast they saw between his domination of the chemical profession up to the 1840s and the destruction of his elaborate system during that decade. To them, Berzelius seemed to bridge two periods of upheaval, the revolution centering on Antoine-Laurent Lavoisier (1743–1794) and the shift from Berzelian chemistry to the new kind of research taking place in the increasingly predominant organic chemistry. Berzelius was readily viewed as the consolidator of revolutionary gains, not an innovator; as an exponent of what had become the exhausted traditions immediately antedating the era to which the nineteenth-century authors belonged. To them, Berzelius was part of an outmoded past; he had dominated his time only because he was its best spokesman; but now, he and his era were fading. His only enduring achievement was the vast accumulation of empirical data that had underlain the theory of chemical proportions.[4]

Accordingly, traditional authors dissected Berzelius's well-known system with two aims: Thinking of it as chiefly a well-integrated articulation of prevailing professional commitments, they could exhibit how its pieces epitomized the chemistry practiced by the generation that followed Lavoisier; and by thus exposing the central motifs of the chemistry that preceded their own, they could lay bare the points of contention that distinguished Berzelius and his generation from theirs. The fall of the Berzelian system in the 1840s could thus be portrayed as another revolution in which new forces overturned the beliefs represented by the central and characteristic ideas of Berzelius's chemistry. The image of an achievement having significance for only one generation is reinforced in

3 *Essai sur la théorie des proportions chimiques et de l'influence chimique de l'électricité* (Paris, 1819). This volume was a slightly revised version of the third volume (1818) of Berzelius's Swedish-language *Lärbok i kemien*, 3 vols. (Stockholm, 1808–1818), together with tables of atomic weights, formulas, and formula weights, and analytical results, originally published separately in 1818 as a supplement to the third volume of *Lärbok: Tabell som utvisar vigten of större delen vid den oorganiska kemiens studium märkvärdiga enkla och sammansatta kroppars atomer, jemte deras sammansättning, räknad i procent: Bihang till tredje delen of Lärboken i Kemien* (Stockholm, 1818). See also Colin A. Russell, "Introduction to the reprint edition [of the *Proportions chimiques*]," in Berzelius, *Essai sur la théorie des proportions chimiques* ... The Sources of Science, No. 99 (New York, 1972), pp. [v]–[xlix].

4 For example, Kopp, *Entwicklung*, pp. 312–314, 582–631; Söderbaum, *Werden und Wachsen*, pp. 92–97, 132–135; Meyer, *Geschichte*, pp. 212–230.

these narratives by their emphasis on the insistent conservatism that Berzelius is said to have exhibited in his last years. Its author inflexible, its leading tenets felled by telling blows, the system of Berzelius inevitably crumbled to make way for new developments.

This view underestimates Berzelius for several reasons. First, it erroneously supposes that original and creative science is confined to instances of revolutionary change. However, as Kuhn suggests, the great bulk of scientific practice is nonrevolutionary, tradition-bound, "normal science." Those who become the partisans of revolutionary innovations need all the skill, creativity, and resourcefulness they can muster if they are to articulate revolutionary achievements and extract from them new kinds of results. Second, by directing attention to tradition in Berzelius's work, the older image of Berzelius ipso facto diverts historians from the novelties. As is discussed here, some of Berzelius's contributions depended on his departing significantly from just those aspects of Lavoisier's chemistry that were *nonrevolutionary*. Third, as Anders Lundgren and Alan Rocke have argued, the traditional picture suffers from linguistic confusions that have blinded not merely historians but also Berzelius's contemporaries to his differences with Dalton.[5] Much more of nineteenth-century stoichiometry issued from the laboratory and pen of Berzelius independently of Dalton and much less of what he did coincided with Dalton's achievements than historians typically have suspected.

Moreover, the traditional accounts do not agree with one another in their choice of the supposedly characteristic features of Berzelius's chemistry and therefore also about the nature of the distinctions between Berzelius and their own generation. These divergent assessments offered by traditional authors likely stem from their procedure of beginning with the fully articulated system of Berzelius and trying to exhibit it as the actualization of antecedent novelties. A more enlightening approach would be to inquire into the functions that Berzelius devised his system to perform the goals Berzelius set for himself, and how well he supposed he had met them. Only by dropping the vantage point provided by hindsight and attempting instead to recover intact the perspective of a figure from the past can the historian assess the originality of a scientist or understand the kinds of distinctions that separated a scientist (like the young Berzelius) from his predecessors or (like the aging Berzelius) from his successors.

This chapter elaborates the perception that Berzelius's chief preoccu-

---

5 Lundgren, *Berzelius och den kemiska atomteorin*, p. 126; Alan Rocke, *Chemical Atomism in the Nineteenth Century: From Dalton to Cannizzaro* (Columbus, OH, 1984), pp. 10–15.

pation was the discrimination of species of matter (compounds from one another and from mixtures) and that the central function of his system was to serve this goal. To this end, Berzelius invoked a combination of qualitative rules, stoichiometric laws, and interpretative procedures for assigning formulas from analytical data. These efforts allowed him to place the entire science of chemistry on a rigorous stoichiometric foundation, something Dalton had called for but scarcely achieved. Moreover, Berzelius attempted to extend his methods to both mineralogy (the locus of much traditional Swedish effort to discriminate different kinds of matter) and to organic chemistry (which, like much of mineralogy, was still generally pursued by natural-historical methods). Though his achievements in organic chemistry were less successful (at least up to the *Proportions chimiques*) than those in mineral chemistry and mineralogy, Berzelius nevertheless taught the chemical profession, as no one had before, that in organic chemistry as in mineral chemistry, the distinct compound in definite proportions must be the focus of chemical practice.

The Problem of Specificity

The origins of Berzelius's preoccupation with specificity are not hard to find. In a general way, Berzelius merely pursued what had been an overarching preoccupation with taxonomy that chemistry in the eighteenth century shared with other branches of science; and in part he carried on a characteristically Swedish aspect of that preoccupation, the chemical classification of mineral species. Botany was the archetype of the taxonomic approach to science in the eighteenth century, and its natural-historical methods and presuppositions were broadly characteristic of taxonomists in other fields. Chemistry shared the less technical features of botany and natural history generally, and Berzelius may be understood as carrying them forward. Moreover, the philosophical foundations and the more technical and practical requirements of botanical taxonomy could be felt in other branches of natural history, such as mineralogy. In this field, too, Berzelius carried forward long-standing preoccupations of practitioners. However, in both chemistry in general and mineralogy in particular he fundamentally altered prevailing classifications and therefore the central preoccupations of research.

Species in Mineral and Organic Chemistry

Eighteenth-century chemical taxonomies invoked the same sorts of easily accessible, qualitative characteristics exploited by botanists to classify plants. This approach fitted well with the assumption, typical of the age, that the properties of compound substances were owing to the property-bearing principles of which they were thought to be composed. For example, both Georg Ernst Stahl (1660–1734) and Herman Boerhaave (1668–1738), two leading figures of eighteenth-century chemistry, explained combustibility by supposing the presence in combustibles of a principle of combustibility (*phlogiston* and the *pabulum ignis*, respectively). G. F. Rouelle (1703–1770), another prominent chemist, classified neutral salts as compounds of an acid with any base capable of coagulating it into a typically crystalline solid. French followers of Stahl, such as L. B. Guyton de Morveau (1737–1816), classified metals as compounds of metallic earths and phlogiston. Lavoisier classified acids as compounds of oxygen with radicals and gases as compounds of the caloric fluid (the reification of heat) with various material substrates. In all cases, membership in a group of substances (a genus) resulted from the presence in a member of the group (a species) of the substance bearing the generic character. Hélène Metzger characterizes these classifications as two-component theories.[6]

The viability of two-component theories rested on three conditions: the ability to point to the loss of the generic property with loss of the generic principle (such as loss of combustibility or metallicity with loss of phlogiston),[7] the ability to demonstrate directly the existence of the specific component after the generic component had left it, and the ability to show that the specific component of each member of the genus indeed differed from those of all the other members (calcination of different metals, for example, left behind demonstrably different calces). The proliferation of two-component theories in the second half of the eighteenth century depended on advances of analysis that revealed the existence of new collections of specific chemicals (like the acids of Lavoisier's oxygen theory or the metallic calces invoked by Guyton).

The irony in this development lay in the generic bias of two-component theories. Despite their dependence on advances in analysis, they characteristically placed great emphasis on the generic component and generic properties at the expense of the specific. Metzger's suggestive use of Aristotelian language in this connection can clarify the point: From the standpoint of the genus, the specific component was a "primitive matter" and the generic components were "forms." By combining with the specific component, the generic ingredient informed it, imposing on it the generic properties. Eighteenth-century chemists, especially in France, were

---

6 Melhado, *Berzelius*, pp. 37–45, 88–99; Hélène Metzger, *Les concepts scientifiques* (Paris, 1926), pp. 56–64; Metzger, *Newton, Stahl, Boerhaave et la doctrine chimique* (Paris, 1930), especially pp. 158–188, 228–245.

7 The property-bearing principles were typically thought to exist only in combination with other bodies; the chemist discerned their existence by reflecting on the events of reaction. Only the specific components, not the generic principles, were thought to be isolable.

vastly more interested in the generic principles and the properties they imposed than on the specific components. Rouelle, for example, characterized salts as "formed by the combination of an acid with any substance which serves as its base and gives it a concrete or solid form."[8] Essential to the genus of neutral salt was an acid, but the specific components were of little consequence. Accordingly, for Rouelle, there could be no question of the "double-entry two-component theory" that Metzger observes in a more modern form of chemistry. A double-entry theory may classify salts by either acid or base because the two kinds of substances are ontologically equivalent. For Rouelle, and indeed all authors of two-component theories, this equivalence was absent. Like the Aristotelian primitive matter, a base (as its name implies) was a passive substrate for the coagulation of an acid into a salt. There was no collection of positively defined basic properties. In his textbook, Guyton broadened the use of the word base beyond the saline theory and invoked it to refer to the specific components of any genus.[9]

The generic bias of French chemistry in the eighteenth century also marked Lavoisier and survived the chemical revolution. Lavoisier was more interested in oxygen than in acidifiable bases; in acids than in salifiable bases; in the caloric fluid than in the bases it rendered elastic; and generally in generic principles than in specific chemicals.[10] A significant consequence of this generic bias was Lavoisier's failure to devise a suitable classification for metallic oxides. Though oxygen was the acidifying principle, higher oxides of metals often neither displayed acidic properties nor served as substrates for salification. As is argued here, one of Berzelius's innovations was to break down the generic bias of chemical classifications and define bases positively. The result was to refer the properties of compounds not to any generic components, but to their constituent species. In his scheme, metallic oxides possessed equivalent status to the nonmetallic.

Despite the proliferation of two-component theories in the eighteenth century, they did not suffice to explain an increasingly common observation: that more than one compound could be formed by the same constituents. Differences in the proportions of the ingredients were readily invoked in this case, however, giving a new career to the "diverse-proportion theories" that had antedated the two-component theories. Stahl's theories of salts and metals, for example, had accounted for both

8 Partington, *History*, Vol. III, p. 74.
9 Guyton de Morveau, Louis Bernard, Hughes Maret, and J. F. Durande, *Elémens de chymie, théorique et pratique, rédigés dans un nouvel ordre, d'après les découvertes modernes, pour servir aux cours publics de l'académie de Dijon*, 3 vols. (Dijon, 1777–1778).
10 Melhado, *Berzelius*, pp. 71–82.

the generic and the specific properties of these bodies in terms of differing proportions of the principles common to all members of the genus. In both cases, water and three kinds of earth constituted all species, which differed by proportions. The advance of analysis, by suggesting the existence of a collection of specific components (in the case of metals, for example, the metallic earths or calces), fostered the replacement of diverse-proportion theories by two-component theories. However, because two-component theories could not account for such cases as the multiple oxides of metals, figures like Lavoisier invoked diverse proportions. In doing so, Lavoisier tacitly assumed definite proportions, and, in the case of oxides he combined his diverse-proportion theory with his two-component theory of acidity: the extent to which an oxide exhibited acidity depended on the (definite) quantity of oxygen it contained. That most other chemists had shared this belief in definite proportions without explicitly articulating it became clear from their reactions to the Proust-Berthollet controversy, in which Lavoisier's colleague Claude Louis Berthollet (1748–1822) ably but ultimately unsuccessfully defended indefinite composition. Berzelius, accepting definite proportions, later gave Lavoisier's correlation of oxidation and properties a thoroughgoing reconstruction.

In organic chemistry (or, more accurately, vegetable chemistry and animal chemistry), the qualitative approach also prevailed, but less agreement existed about the results of its application. In successive steps, fire analysis (exposure of organic matter to naked distillation) and, more often in the late eighteenth century, extractive analysis were used to separate vegetable or animal matter into collections of substances such as gels, gums, oils, resins, and starches. Any taxonomy of distillates or extracts was unstable, however, for the products of analysis were always subject to further analysis and increasingly were resolved into additional products. By the opening of the nineteenth century, the five principles dating from the iatrochemists of the sixteenth century had become the twenty categories of vegetable substances listed by Antoine François de Fourcroy (1755–1809).[11]

At first, these categories paralleled those of the botanical and zoological taxonomies by referring substances to the plant or animal of origin. However, with the advance of analytical techniques, discrimination of these substances was achieved by invoking several new sorts of characters, including the analytical techniques by which they were obtained and the reactivities and other notable properties of the products of analysis. Fourcroy's *Connaissances chimiques* (1800–1802) is taken to be a wa-

---

11 Antoine François de Fourcroy, *Système des connaissances chimiques*, 11 vols. (Paris, 1801–1802), Vol. VII, pp. 51–56, 111–113, 120–127.

tershed in this connection. There he replaced the older botanical taxa with chemically characterized categories of the "immediate principles" or, as he later called them, the "immediate materials," that is, the supposed ingredients of plants that he had obtained by largely extractive measures.

Substances with sharply differing qualitative characters were readily taken to possess significant differences in composition, but substances with slight qualitative distinctions occasioned disagreement. Lavoisier, for example, calling attention to the few vegetable oxides known in the late eighteenth century, cited one genus of similar substances and two others that he did not label as genera: "The vegetable oxides with two bases [carbon and hydrogen] are sugar, the different species of gums that we had united under the name of mucous bodies, and starch." He then described "these three substances" as if the members of the genus of gums might just as well be considered the same substance, for the specific distinctions were not pronounced.[12] Another example is the division of opinion around the turn of the century about the specificity of the organic acids. Originally, the acid of vinegar was the sole organic acid clearly recognized as a distinct substance, but Carl Wilhelm Scheele (1742–1786) had discovered several new ones in the 1780s. Though he considered them distinct substances, other chemists claimed that these new substances were merely modifications of acetic acid (which they closely resembled). Still others invoked the modification theory for some organic acids but not others. Fourcroy and Louis Nicolas Vauquelin (1763–1829) considered malic acid distinct, but others, particularly those (like lactic acid) that do not crystallize, as modifications (perhaps resulting from impurities). Lavoisier, however, listed all the organic acids as distinct substances. Their characters were sufficiently distinct to permit this judgment, but not sufficiently so to compel it.[13]

What distinguished, in the minds of contemporaries, the cases in which differences in properties were too small to signify species boundaries from those in which the boundaries were thought clearer? In some cases, as just noted, practioners cited impurities. In others, however, it is quite possible that the law of definite proportions was not assumed for organic matter. In the light of the overall uniformity that eighteenth-century analysts perceived in organic matter, what could justify deferring to a minuscule difference in the percentage composition of two gums, for ex-

---

12 Antoine Laurent Lavoisier, *Traité élémentaire de chimie*, 2 vols. (Paris, 1789), Vol. I, pp. 125–126.

13 Hugo Olsson, *Kemiens historia i Sverige intill år 1800*, Lychnos Bibliotek, Vol. 17 (Uppsala, 1971), pp. 351–360; Maurice P. Crosland, "Lavoisier's Theory of Acidity," *Isis*, 1973, 64:306–317; E. Hjelt, *Geschichte der organischen Chemie* (Brunswick, 1916), pp. 10–11; Partington, *History*, Vol. III, p. 555.

ample, to justify assigning two eminently similar substances to different species? In mineral chemistry, species boundaries between different substances composed of the same ingredients involved jumps in composition; in organic chemistry, clear jumps were absent. Doubtless contemporaries considered the occurrence of transitional forms more likely. This conflict between definite and indefinite proportions, between saltation and transition, also showed itself very distinctly in eighteenth-century mineralogy.

Species in Mineralogy

As a traditional branch of natural history, mineralogy in the eighteenth century reflected more clearly than mineral and organic chemistry the formal and technical preoccupations of botany, which are epitomized in the work of Carl von Linné (1707–1778), the leading naturalist of the age. His work shares the aspiration of virtually all naturalists of the period to devise a "natural system" of classification, that is, one that would express the divine plan of nature, successfully discriminating various taxa while exhibiting the transitions (gradual shifts in the values of their characters) among them. Recognizing the difficulty in reaching that goal, naturalists resigned themselves to producing "artificial methods" instead, that is, those that could classify nature but did not actually express its plan of organization. Though they thus tabled their major goal, practitioners nevertheless judged their artificial systems by the extent to which they seemed able to approximate a natural method. Linné, for example, chose to classify according to characters drawn from the "fructification" of plants, their flower and fruit, for he considered these organs as somehow essential elements of their being and reflective of their nature. His well-known sexual system drew its legitimacy from the expectation that its taxa and the transitions among them closely approximated those of the natural method, an expectation thought justified by the correspondence that Linné established between the sexually defined genera and a set of natural genera that he had devised by reflecting on his perceptions of similarity.[14]

In mineralogy, followers of Abraham Gottlob Werner (1749–1817) emulated the Linnaean example most closely.[15] The Wernerian school, preoccupied by transitions, invoked as taxonomically significant charac-

14 Julius von Sachs, *History of Botany (1530–1860)* (Oxford, 1890); James L. Larson, *Reason and Experience: The Representation of the Natural Order in the Work of Carl von Linné* (Berkeley, 1971).

15 For the mineralogical traditions described in the next several paragraphs, see Melhado, "Mineralogy and the Autonomy of Chemistry around 1800"; and Melhado, "Mitscherlich's Discovery of Isomorphism."

ters only those in which transitions could be observed. Their justification was the belief that these characters must reflect the essence of minerals, just as the fructification of plants reflected their essential nature. Like virtually all mineralogists in the century, Wernerians believed that essence lay in material composition, and they maintained, consistently with Werner's geological ideas, that composition also exhibited gradual transitions. Mineral genesis took place by gradual precipitation, with gradual change in composition, from a primordial sea. The transitions in the external characters that they invoked to construct mineralogical taxa were considered reliable reflections of the transitions in composition. By contrast, chemical analysis was regarded by Wernerians as unreliable. Moreover, the Wernerian position was inconsistent with definite proportions, and it allied the Wernerian school with Berthollet's notions of indefinite composition.

The Wernerian school crystallized in great measure in response to the chemical school of mineralogy, which took precisely the opposite position regarding the relation between material essence and external characters: The latter were unreliable reflections of essence, which could be revealed only by chemical analysis. Before Berzelius, however, this school did not invoke definite proportions, and it suffered from the very weakness most often cited by the Wernerians, the rudimentary state of chemical analysis. The school took form in Sweden, descending from Johan Gottschalk Wallerius (1709–1785) and Axel Fredrik Cronstedt (1722–1765), and found its fullest expression in the work of Torbern Bergman (1735–1784). To this Swedish tradition may very likely be traced the great interest Berzelius displayed in the specific diversity of matter (in contrast to the French preoccupation with generic groups). Bergman cited both practical and theoretical reasons for preferring the chemical approach: The industrial motivations of Swedish mineralogy put a premium on knowledge of the content of minerals and the correlation between composition and properties was unreliable. Bergman found in the traditions of blowpipe analysis and the more recent advances in wet-chemical methods (to which he made significant contributions) as grounds for justifying a chemical approach. One of the leading chemists of the age thus articulated a position about mineralogy (external characters do not reliably reflect essence) that lay at odds with the traditional belief of chemists that properties reflect composition (i.e., that the properties of substances reflect the property-bearing principles of which they were composed). The change shows how the practice of laboratory analysis displaced the method of mental analysis (deciding on composition by reflecting on properties).

In his taxonomy, Bergman identified a group of substances that he claimed were the proximate constituents (in practice, elements) of all minerals: earths, acids, alkalies, metals, and phlogistic bodies. He described minerals as chemical compounds of these substances and constructed genera and species solely with reference to composition. Each genus took its name from one of these proximate constituents and included all species that contained a larger proportion by weight of the generic constituent than of other constituents.

This method of classification paid virtually no attention to the external characters and natural affinities preoccupying the Wernerians. It also introduced into mineralogy a variety of new substances and tended to erase the boundaries between mineralogy and chemistry. Whereas the naturalist excluded from consideration any artificial products (whether obtained by analysis or synthesis) that did not also occur naturally (for they lay outside the natural order that the naturalist aimed to depict), the chemical mineralogist found little reason to distinguish them. After some hesitations based on his deference to natural-historical tradition, Bergman decided to include in his scheme both the proximate constituents (even if they did not occur in nature in an uncombined state) and a vast variety of salts (far beyond the simple salts that later claimed the bulk of Lavoisier's attention), many of which were known only in the laboratory. Bergman's well-known reform of the chemical nomenclature drew on his mineralogical taxonomy; indeed, he saw no basis for distinguishing mineralogy from chemistry. In this respect, Berzelius can be understood as a successor to Bergman. Berzelius's system of nomenclature of 1811 similarly obscured the boundaries of chemistry and mineralogy, and his later effort to deploy his system in mineralogy was a conscious attempt to subsume mineralogy under mineral chemistry.[16] In at least one significant respect, however, Bergman's mineralogy lacked novel implications: Because it did not invoke definite proportions, it posed no serious challenge to the Wernerian preoccupation with transitions.

The crystallographic tradition in mineralogy, however, did postulate definite proportions, and Berzelius drew inspiration from it in his struggle with the natural-historical approach. Particularly as developed by René Just Haüy (1743–1822), the crystallographic tradition united a chemical species definition with morphological analysis and thus formed a middle ground between chemical and Wernerian approaches. For Haüy, the specificity of a mineral was owing to the specificity of the form of its "integrant molecules" (the smallest bits of a substance possessing its characteristics); specificity of form, in turn, depended on the definite pro-

16 Berzelius, "Essai sur la nomenclature chimique," *Journal de physique*, 1811, 73:253–286; Berzelius, *Attempt to Establish a Pure Scientific System of Mineralogy*, tr. John Black (London, 1814). The German translation of this work and of a second, closely related treatise are discussed near the end of this chapter.

portions in which the constituents of the mineral had entered into chemical composition. Like his predecessor, J. B. de Romé de l'Isle (1736–1790), Haüy had postulated definite proportions as the foundation of his mineralogy. All samples having the same integrant molecule and the same composition belonged to the same species. Haüy anticipated a division of mineralogical labor, with chemistry fixing minerals in the taxonomic system and morphological analysis determining which samples of diverse external appearance belonged in the same species.

Haüy's mineralogy, like Bergman's, violated Wernerian species boundaries. His morphological methods allowed him to redraw species boundaries, even if morphologically identical samples varied in composition. Definite proportions, however, demanded that Haüy distinguish essential ingredients (those combined in definite proportions) from accidental ones (impurities mixed together with the essential substances). Variability in the accidental ingredients produced the variations in external characters that Wernerians used to determine species. As reflections of accidents, however, the external characters of the Wernerians signified nothing fundamental for Haüy. Around the turn of the century, however, practioners of both chemistry and mineralogy became increasingly aware that morphologically identical samples possessed great variability in composition. Faced with such evidence (pressed, for example, by Berthollet, champion of indefinite composition), Haüy admitted that chemical analysis could not perform the task for which he invoked it (fixing the chemical identity of minerals), and he was compelled to resort to crystal morphology, hoping to exploit it as a control over the results of chemical analysis. Berzelius hoped to accomplish what Haüy could not: chemical determination of species. With suitably articulated stoichiometric laws, he anticipated, all questions about accidental ingredients would be solved.

The Discrimination of Kinds: Electrochemistry and Stoichiometry

As a device for discriminating and classifying compounds, Berzelius's system jointly invoked two clusters of ideas, one qualitative and one quantitative. The qualitative cluster, which derived largely from Berzelius's early studies on the chemical effects of the Voltaic pile, constituted the locus of many of his departures from Lavoisier. The quantitative side issued from his early interest in pre-Daltonian stoichiometry. Throughout his career, Berzelius kept the qualitative and quantitative sides of his work in juxtaposition, insisting that a compound can be characterized fully only by joint exhibition of its quantitative composition and its fundamental qualitative features.

The Electrochemical Sorting Phenomenon

Those fundamental features were electrochemical in nature, in particular the possession by compounds of ontologically equivalent but electrochemically opposed constituents. The generically biased two-component theories with which Lavoisier had described acids and salts gave way to a double-entry two-component theory in which the basic components were no longer passive substrates but coequal partners. Any substance that could be shown to possess such opposing constituents could be described as saline and made subject to the quantitative laws that Berzelius devised to express the composition of salts.

Though the successful decomposition of the alkalies and earths gave the decisive impetus to this double-entry theory, the way had been prepared by Berzelius's prior studies of the electrochemical sorting phenomenon. Collaborating with Wilhelm Hisinger (1766–1852), a wealthy industrialist who befriended Berzelius and patronized his career, in a paper of 1803 Berzelius exhibited the ability of the newly invented Voltaic pile or battery to separate salts into constituent acids and bases, which were then sorted by accumulation around the opposite poles.[17] Their conclusion suggested that the sorting phenomenon implied the ontological equivalence of the constituents. Also emerging in the 1803 paper was an electrochemical generalization of the notion of acidity and basicity. Berzelius and Hisinger pointed to the separation by the pile of the saline constituents into their constituents – typically radicals and oxygen – and suggested what was to become one of the fundamental Berzelian tenets: All substances were in principle subject to separation by the pile and that those associated with a given pole possessed a certain analogy among themselves. The archetypes for each group were the typical acids and bases, but substances not typically regarded as acidic or basic were analogous electrochemically to ordinary acids or bases, respectively. Accordingly, acidity and basicity as usually understood were mere extremes of more general properties, electronegativity and electropositivity.

These implications of the study with Hisinger were decisively reinforced by the electrochemical decomposition of the alkalies and earths by Humphry Davy (1778–1829) and Berzelius. Following Davy's successful decomposition of the alkalies, Berzelius collaborated with Magnus Martin Pontin (1781–1858) in a study leading to the decomposition of potash and several earths into metals and oxygen, the discovery of ammonium amalgam, and the suggestion that ammonia, like other alka-

---

17 Berzelius and Wilhelm Hisinger, "Versuch, betreffend die Wirkung der elektrischen Säule auf Salze und auf einige von ihren Basen," *Neues allgemeines Journal für Chemie,* 1803, 8:269–324.

lies, was an oxide, in this case of the amalgam, a compound radical.[18] Three consequences flowed from this demonstration that the typical bases, like the typical acids, were oxides. First, it accorded to the metallic oxides, misfits in Lavoisier's chemistry, a positive place in Berzelius's. This theoretical insight was made largely explicit in Berzelius's paper of 1811 on nomenclature and was realized experimentally in a programmatic study of oxides and sulfides (discussed later). In the paper of 1811, Berzelius pressed the view that higher metallic oxides could form salifiable bases, possessed positively enumerable properties opposed to those of acids, and therefore required denominations that Lavoisier's nomenclature lacked (Berzelius thus applying to them the "-ous/-ic" nomenclature previously confined to the acids). Second, it precluded two-component theories of acidity and basicity (though not before both he and Davy had toyed with a generic principle of alkalinity). The conclusion eventually seemed compelling that the seat of acidity or basicity must lay in the radical. What had been the generic properties of oxides were henceforth to be seen as the specific properties of radicals.

Third, by reinforcing the notion of the ontological equivalence of acids and bases, Berzelius's discovery of the oxide nature of the bases led him to pursue the insights articulated in the paper with Hisinger and to construct a new vocabulary suitable to express them (though only in his *Proportions chimiques* of 1819 did he attain his mature position about these themes).[19] Acids and bases were ontologically equivalent, all of them were oxides (or sulfides) of radicals, acidity or basicity resided in the radicals, and all radicals possessed acidity or basicity to at least some degree (i.e., could be understood as relatively acidic or basic in comparison with other radicals). The typical acids and bases were therefore just extremes on a continuum, and the degree of oxidation (which Lavoisier considered the sole determinant of acidity) must serve only to modify the expression of the inherent acidity or basicity. In the *Proportions chimiques,* Berzelius offered an electrochemical series of radicals, that is, a smooth scale, running from strong electronegativity (or acidity) through strong electropositivity (or basicity). The order of radicals in the series could be determined by comparing the oxides, and the effects on the order of varying degree of oxidation could be largely eliminated by comparing radicals at a single level of oxidation (the highest).[20] The radicals

18 Berzelius and Magnus Martin Pontin, "Electrisch-chemische Versuche mit der Zerlegung der Alkalien und Erden," *Annalen der Physik,* 1810, 36:247–280.

19 Moreover, they were developed in parallel with the experimental work, described below, that underlay Berzelius's reconstruction of Lavoisier's correlation of acidity and basicity.

20 Berzelius, *Proportions chimiques,* pp. 90–91. However, his progress toward this recognition was hindered by the weight of precedents in Lavoisier's chemistry and his own

had come into their own; gone were most vestiges of Lavoisier's two-component theory of acidity.

The *"Versuch"* and the Orders of Composition

These insights into the equivalence of acids and bases and the role of radicals in determining properties did not shift Berzelius's attention back from oxides to the radicals per se. For Berzelius, the characteristic properties of the radicals were displayed by their oxides but were largely remitted in the direct combinations of radicals with one another. The electrochemical series thus depended on Berzelius's studies of the qualitative characteristics of oxides. However, the modifications in electrochemical character correlated with degree of oxidation, a quantitative factor. As will be noted shortly, Berzelius's interest in quantitative analysis was independently motivated by his knowledge of pre-Daltonian stoichiometry. The confluence of this concern with the sequelae of the early electrochemical studies stamped the bulk of Berzelius's experimental work. He combined qualitative investigation of the electrochemically defined properties of oxides with quantitative determination of their composition (and eventually that of their salts, conceived as compounds of oxides with one another). His early experimental investigations of oxides and salts were reported in his "Versuch, die bestimmten und einfachen Verhältnissen aufzufinden, nach welchen die Bestandtheile der unorganischen Natur mit einander verbunden sind" [Attempt to discover the definite and simple proportions in which the constituents of inorganic nature are combined] (1809–1812).[21]

Berzelius's knowledge of stoichiometry was one outcome of extensive reading he had undertaken in the early years of the century to prepare an up-to-date Swedish chemistry text.[22] From Jeremias Benjamin Richter (1762–1807), whose works were even then obscure, Berzelius obtained two laws governing the composition of salts, the law of neutrality and the law of equivalent proportions. He independently discovered a third law that, it turned out, Richter had also articulated as a consequence of the law of neutrality, called here the "law of the constancy of basic oxy-

difficulty in distinguishing the role of radicals from that of oxygen in determining acidity or basicity; see Melhado, *Berzelius,* pp. 159–163.

21 Republished as Berzelius, *Versuch, die bestimmten und einfachen Verhältnisse aufzufinden, nach welchen die Bestandtheile der unorganischen Natur mit einander verbunden sind,* ed. Wilhelm Ostwald. Ostwald's Klassiker der exakten Wissenschaften, No. 35 (Leipzig, 1892), from which citations are taken; see Melhado, *Berzelius,* pp. 168n, 181n, 202n–203n.

22 Söderbaum, *Werden und Wachsen,* p. 76; Berzelius, *Autobiographical Notes,* ed. H. G. Söderbaum, tr. Olof Larsell (Baltimore, 1934), pp. 65, 140–141; Lundgren, *Berzelius och atomteorin,* pp. 72, 75–76, 91, 97.

gen," and still another, unprecedented law, called here the "rule of oxides." The law of neutrality, a special case of equivalent proportions, concerned the phenomenon of maintenance of neutrality upon mixing two solutions, each containing a different neutral salt. If the combining proportions of the constituents of each of the two salts that were to be mixed were known, the proportions of the salts produced in the double decomposition that ensued upon mixing could be calculated directly. More generally, the law of equivalent proportions states (in more modern terms) that the amounts of the bases that saturate a given amount of an acid always stand in the same proportion to one another; and that the same holds for the amounts of the acids that saturate a given base. Berzelius's early analytical work was motivated in part by his effort to exploit Richter's law of equivalent proportions in calculating the composition of a large number of salts from a small body of analytical data. The failure of available analytical results to conform to Richter's law induced Berzelius to perform his own analyses of salts, studies that led to the two additional laws, the constancy of basic oxygen and the rule of oxides. These laws linked the composition of oxides with that of their salts. The constancy of basic oxygen stated that the oxygen in the bases of all the neutral salts of an acid was the same; the rule of oxides asserted that the basic oxygen of a salt is always a small integral submultiple of the acidic oxygen.

Berzelius's interest in quantitative studies thus had already been sparked by the time of Davy's decomposition of the alkalies. The electrolytic decomposition of hitherto undecomposed substances offered new material for submission to quantitative study and calculation. Ammonia in particular assumed in this connection a special importance for Berzelius. Regarding it as an oxide of a metallic substance manifested in mercury amalgam, he hoped to determine its oxygen content. Because of Berzelius's repeated failures to isolate the substance and because of its supposed rapid oxidation even when amalgamated, the oxygen content could be determined only by calculation with Richter's laws. To this end, Berzelius needed exact analyses of a variety of metallic oxides, at least one acid, and the salts formed by these constituents. From this ambition emerged the "Versuch."

The dual background of the "Versuch" (the concern for qualitative, electrochemical characterization and the desire to calculate composition) had directed Berzelius's attention to two levels of composition that he later described as being of the first and second order (compounds of radicals with oxygen [and, later, with sulfur] and salts (the compounds of oxides or sulfides with one another). Both the law of basic oxygen and the newly decomposed bases attracted Berzelius to the problem of determining the composition of oxides. Investigation of the composition of first-order oxides, and sulfides, was a major part of Berzelius's long effort to establish combining weights and the oxidation and sulfuration stages of all elementary radicals and provided the material for the reconstruction of Lavoisier's correlation of oxidation and properties. On the other hand, the laws of neutrality, of equivalent proportions, of basic oxygen, and the rule of oxides all applied to salts. Unlike Dalton, Berzelius always pursued his quantitative studies with simultaneous reference to first- and second-order composition. The discovery and articulation of laws connecting orders of composition was a central technique in this enterprise; indeed, the rule of oxides and its progeny became cornerstones of Berzelius's system.

The Rule of Oxides

First announced in a letter to Ludwig Wilhelm Gilbert (1769–1824) of 1 October 1810, the rule was given its earliest systematic elaborations in the first of three "Continuations" of the "Versuch." There, Berzelius described it thus:

> In neutral salts the amount of oxygen contained by the acid is an integral multiple of the amount of oxygen in the base. This rule can be stated a bit more generally and, I believe, no less correctly as follows: When two oxidized bodies saturate one another, they always contain oxygen according to such proportions that the amount of oxygen in the body that goes to the positive pole in the circuit of the electrical pile is an integral multiple of the amount of oxygen in the other body, the one that tends toward the negative pole.[23]

The rule was the outcome of a chain of reasoning and experimenting that began with the law of multiple proportions (which Berzelius reached either through independent discovery or through the literature).[24] Multiple proportions led Berzelius to recalculate analyses of oxides of lead and sulfides of iron to reveal not percentage composition but the multiple proportions in which a given weight of one substance combined with the other. Having exhibited series of lead sulfides and iron oxides, he tried to find a similar pattern of multiple proportions in the sulfides of lead. However, he could produce only one compound of these elements. He therefore attempted to determine the proportions that would be exhibited by other potential compounds of these substances by inquiring whether,

---

23 Berzelus, *Versuch*, p. 88 (this, and all other translations, unless otherwise noted, are those of the author).
24 Melhado, *Berzelius*, pp. 170–184.

when oxidized, the radicals of lead and sulfur combined in the same proportions as when unoxidized (that is, whether the same proportions of lead and sulfur existed in both lead sulfate and lead sulfide). Berzelius's proof that the proportion was preserved in the oxidized compound led him to articulate a broad principle, the preservation of combining proportions. From that principle and from additional analytical results, Berzelius was able to argue that the amount of oxygen in the base of any neutral sulfate was the same; and from further studies with muriates Berzelius concluded that the oxygen in the bases of all neutral salts of an acid is the same – that is, he reached the law of the constancy of basic oxygen.

This law was a breakthrough, for it allowed Berzelius to calculate the composition of the alkalies and earths in such a way as to link their composition with that of their salts: The basic oxygen in a metallic salt of an acid must equal the basic oxygen in the salts formed by the acids with alkalies and earths; hence the oxygen content of the alkalies and earths could be determined, as Berzelius showed in the second half of the "Versuch." Among these demonstrations was a calculation of the supposed oxygen content of ammonia.[25]

Much of the first half of the "Versuch" was devoted to exploiting the principle of the preservation of combining proportions and to overcoming anomalies to it. Repeatedly, Berzelius compared the composition of the oxides of metals with the composition of their sulfides in minimo, produced their salts from the sulfides by oxidation, and compared the proportions of the metallic radicals with sulfur in their compounds. Berzelius used this approach to establish the proportions of the oxygen to sulfur in sulfuric and sulfurous acids; to exhibit the proportions of the oxides, sulfides, and sulfates of copper; and to attempt to bring other compounds, such as the sulfides and sulfates of iron, into conformity with the preservation of combining proportions. In the latter case, however, Berzelius found that the salts of the higher oxide of iron failed to display a preservation of the original combining proportions of the sulfide or even an integral multiple of those proportions.

Attempting to overcome this anomaly led Berzelius to the rule of oxides. He noticed that the proportions of oxygen in the two oxides of iron, unlike the corresponding copper compounds, were 1:1½ for a given weight of metal. The conjunction of the sesquialteral proportion with the anomaly to the preservation of combining proportions led Berzelius to suppose that perhaps unknown lower oxides would prove the sesquialteral proportion merely apparent and simultaneously remove the anomaly to the preservation of combining proportions. Not any concern for an atomic

25 Berzelius, *Versuch*, p. 73.

conception of matter, but an anomaly to a stoichiometric principle occasioned Berzelius's discontent with nonintegral multiples.

Hoping to reduce the anomaly, Berzelius returned to the oxyacids of sulfur, which he had shown to exhibit the sesquialteral proportion in the amount of oxygen for a given weight of radical. He sought to reduce that proportion by finding at least one lower oxide of sulfur. Using the law of equivalent proportions and existing analytical data about the combinations of carbon with oxygen and hydrogen, he calculated that sulfur in its lower oxidation stage had to contain exactly half the weight of oxygen for a given weight of sulfur that his own analyses indicated to exist in sulfurous acid and one-third that existing in sulfuric acid. Comparing these results with data for lead salts, Berzelius noticed that this lower proportion of oxygen in the suspected lower sulfide happened to equal the proportion of oxygen in the lead oxide of the lead salts. Accordingly, the sulfuric acid in the lead salts contained three times the oxygen of the lead oxide. This integral proportion subsisting between the oxygen of the base of a salt and that of the acid was the first example of what Berzelius soon generalized into a law, here called the rule of oxides. Like the earlier principles this one connected two orders of composition.

As the outcome of Berzelius's efforts to find a lower oxide of sulfur that would render the sesquialteral proportion apparent, the rule became associated in his mind with certain speculations that later proved to be its consistent concomitants and that furthermore signaled Berzelius's willingness to apply his model of salts to substances that scarcely possessed saline properties.[26] Having calculated the composition of the suspected lower oxide of sulfur, Berzelius hoped to demonstrate its existence. He supposed it unlikely to exist as an isolable substance, instead anticipating its appearance as an ingredient in compounds, preserved there by the "intermediation" of the other ingredients. Berzelius believed he had found both the suspected lower oxide and a second still lower one in two substances (later understood to be two chlorides of sulfur) formed from the combination of sulfur with muriatic acid (itself believed to be the oxide of a radical). Though they lacked saline properties, Berzelius regarded these substances as salts in which the oxides of sulfur played the basic role in combination with muriatic acid.

The intermediation of the muriatic acid permitted the lower sulfides to exist as ingredients of the compounds. Berzelius had earlier used a different form of the idea of intermediation to account for anomalies to the preservation of combining proportions. Now he supposed instead that by permitting the existence in combination of lower oxidation stages than

26 These speculations appeared in later additions to the "Versuch"; they occupy pp. 47–53, 82–87 of the Ostwalds Klassiker edition; see Melhado, *Berzelius*, pp. 180–190.

can exist in isolation, intermediation could eliminate the sesquialteral proportion. He speculated that "perhaps combinations of this kind occasion multiples by 1:1½, so that the latter always presuppose lower oxidation stages by which they are multiples by 6 or 12. And perhaps in the future it will be shown that these stages are always increments by the even numbers, 2, 4, 6, 8 and perhaps higher."[27] Berzelius now anticipated that this pattern would prove commonplace in organic chemistry, but in fact it was in his studies of inorganic composition that he searched for lower, hidden oxides in his efforts to exhibit the patterns of oxidation and salt formation.

Experimental Elaboration of the Theory of Salts

Berzelius's discovery of the rule of oxides intensified his commitment to the themes of research that had culminated in it and invested his subsequent studies with a programmatic character. The program comprised three distinct but related branches: the quantitative analysis of simple binary compounds, especially the oxides and sulfides of radicals; the qualitative characterization of oxides and sulfides in terms of the generalized notions of acidity and basicity; and the quantitative analysis of salts, that is, the compounds of oxides and sulfides with one another. The program was novel in part because it constituted the first effort to discover and analyze all the oxides and sulfides of the known radicals. By contrast, the immediate followers of Lavoisier, such as Fourcroy, had attempted to discover new salts, hoping to fill in the blanks in the chart of the *Méthode de nomenclature chimique*[28] exhibiting all possible saline combinations in the belief that all saline constituents, at least the inorganic ones, were known or soon would be. However, a programmatic study of the saline constituents themselves was new. Moreover, many contemporaries, such as Joseph Louis Gay-Lussac (1778–1850), were content to study the compounds of an occasional radical, but virtually none shared Berzelius's ambition to study all of them.[29]

The research program extended and completed the research recounted in the first two major sections of the "Versuch." In the earlier work, Berzelius had attended almost exclusively to the radicals of the bases and the salts they formed with but a few acids. To analyze the bases, particularly in the light of the metallization of ammonia, was a major motive

27 Berzelius, *Versuch*, p. 86.
28 Antoine Laurent Lavoisier, L. B. Guyton de Morveau, Claude-Louis Berthollet, and Antoine François Fourcroy, *Méthode de nomenclature chimique* (Paris, 1787), pl. facing p. 100.
29 The only significant exception is Thomas Thomson, *An Attempt to Establish the First Principles of Chemistry by Experiment*, 2 vols. (London, 1825).

of the "Versuch," and the exploitation of Richter's methods to this end demanded analysis of few acids. Discovering the rule of oxides, however, shifted Berzelius's attention to the acids, because the rule linked the composition of the acidic component with that of the basic one. Quantitative analysis of acids was itself groundbreaking, for, as Berzelius observed in his letter to Gilbert announcing the rule, the composition of many acids was still unknown[30] – this, over 20 years after the publication of Lavoisier's *Traité*, which had stressed the correlation of acidity with composition. Moreover, the rule provided a more powerful means for determining the composition of salts than either preservation of combining proportions or the constancy of basic oxygen. Thus, Berzelius determined the oxidation stages of acidic radicals not only for their own sake, but also to aid in determining the composition of salts.

The study of the acidic radicals was rapidly integrated with the ongoing study of the basic ones. The dividing line between the two, already blurred by the electrochemical generalization of the ideas of acidity and basicity, grew still less determinate as Berzelius increasingly documented what Lavoisier had suspected: that the higher oxides of even metallic radicals might be acidic. From the first "Continuation" of the "Versuch," where Berzelius presented a series of analyses to document the rule of oxides, the fusion of the study of the acidic with the basic radicals became increasingly apparent. The analyses presented there of the compounds of arsenic, a semimetal, provided the bridge between the acidic and the basic radicals and constituted a model for later studies.[31] He gave each radical, regardless of its position in the electrochemical series, similar treatment with similar ends in view.

Throughout this program, Berzelius aspired to reconstruct Lavoisier's correlation of oxidation and properties. His early results, however, thoroughly convoluted what had been the apparently clean correlations of acidity with oxygen content established by Lavoisier.[32] Berzelius nevertheless persisted in seeking to reduce the diversity in the observed patterns of oxidation of the radicals, subjecting them to the constraints that their salts conform, in delimitable ways, to the rule of oxides. Not surprisingly, Berzelius was ultimately compelled to recognize that uniformity in patterns of oxidation and salt formation was not so broad as he would have liked; but, as shown later, he seemed to have succeeded in reducing what had been a great diversity in the stoichiometry of radicals to variations on a few paradigmatic cases.

In the discussion that follows, the content and outcomes of Berzelius's

30 Berzelius, "Schreiben . . . an den Prof. Gilbert. . . . Stockholm, d. 1 Oct. 1810,"
   Annalen der Physik, 1811, *37*:208–220.
31 Berzelius, *Versuch*, pp. 115–121.   32 Ibid., pp. 117–118.

research program are briefly developed. Emphasis is also given to Berzelius's studies of the compounds of nitrogen and the oxides of the acidic radicals and of the weakly reactive metallic radicals. Another weakly reactive oxide, water, also acquired a consequential place in Berzelius's studies and is given commensurate attention. The discussion exhibits the construction of the Berzelian theory of salts and sets the stage for exploring its deployment in mineralogy and organic chemistry.

Lower Oxides and Saline Combinations of Nitrogen

Because Berzelius's generalization of the idea of basicity emerged from his belief in the oxide nature of ammonia, his commitment to that belief was intense. However, various experimental developments suggested that ammonia could not be regarded as an oxide of a compound basis (the amalgam) as he and Davy had at first suspected. Following in Davy's footsteps, Berzelius supposed instead that nitrogen, hydrogen, and ammonia were all oxides of a still unknown radical, which Davy had named "ammonium." Submitting the analytical data to calculation, Berzelius portrayed not only the three gases, hydrogen, nitrogen, and ammonia, as compounds of ammonium, but also the known higher oxides of nitrogen. The oxygen in nitrogen, Berzelius held, contained $1:1\frac{1}{2}$ times the oxygen in ammonia. His earlier speculations about lower oxides having suggested that the sesquialteral multiple merely reflected higher multiples (by 6 or even 12) of lower oxides, perhaps incapable of an independent existence, Berzelius supposed hydrogen was a lower oxide containing one-twelfth the oxygen of nitrogen. It headed a table portraying a series of substances as oxides of ammonium. A suboxide of ammonium followed hydrogen, then came ammonia, nitrogen and its oxides and finally water appeared as the highest known oxide of ammonium. Berzelius had thus accounted for ammonia and a whole series of related substances as a set of simple oxides of a single radical with progressively higher multiples of a minimal combining proportion.[33]

Among Berzelius's arguments for the oxide nature of ammonia and therefore also of nitrogen was that preserving the basic oxygen of ammonia permitted the subsumption of its salts under the rule of oxides. Having calculated the oxygen content of ammonia (in the second half of the "Versuch") with the law of basic oxygen, Berzelius noticed that the composition thus determined conformed to the rule of oxides (that is, in

33 Davy's ideas in this connection may be followed in his *Collected Works*, 9 vols. (1839–1840; rpt., New York, 1972), Vol. V, passim. For Berzelius's views of nitrogen, see Söderbaum, *Werden und Wachsen*, pp. 119–127; Lundgren, *Berzelius och den kemiska atomteorin*, pp. 76–80, 102–113; Berzelius, *Versuch*, pp. 96–110; cf. also Lundgren, Chapter 4 in this volume.

ammonium salts, the acids contained integral multiples of the basic oxy-gen).[34] Berzelius now extended the argument from the opposite direction, that of the nitrogen oxyacids and their salts. Indeed, he prepared a study of neutral nitrates specifically to bolster his contentions about the oxide nature of nitrogen.[35] Two of Berzelius's ideas had thus become mutually reinforcing: The oxide nature of nitrogen was sustained by the conform-ity of nitrates to the rule of oxides, and the rule of oxides was sustained by the reduction of this acid and its salts to the pattern of oxidation and salt formation displayed by sulfur (henceforth called the "sulfur pat-tern"). Berzelius's persistence in upholding the oxide nature of nitrogen must be traced to the gratifying character of this interlocking relation-ship. Berzelius later added studies of nitrites and basic salts of the nitro-gen oxyacids.[36] His work fully established the existence of a series of nitrites, provided a store of firm analyses of the salts of both acids, and contributed to the general question of the composition of basic salts, a major theme discussed here. These results all issued from the initial mo-tivation of strengthening the relationship between the compound nature of nitrogen and the rule of oxides.

However, his view on the neutral salts was flawed by his use of Gay-Lussac's erroneous analysis of nitric acid.[37] With Gay-Lussac's data, Ber-zelius found that nitrates did not conform to the rule of oxides, for the oxygen content of the nitric acid fell between four and five times that in the base. If the oxygen content of the nitrogen in the nitric acid was reckoned into the calculation, however, then the acidic oxygen equaled six times the basic. The rule of oxides thus demanded the compound nature of nitrogen. Eventually, Berzelius discovered the error; but he did not readily give up the compound nature of nitrogen.

Oxidation and Sulfuration: Clarifying the Principle of the Preservation of Combining Proportions

The principle of the preservation of combining proportions asserted the correspondence of the sulfide in minimo with the lowest oxide of a radi-cal and expressed the preservation of the proportion of the basic radical to the sulfur of a sulfide in the corresponding neutral sulfate of the radi-cal. However, in articulating the principle in the "Versuch," Berzelius left unclear the relations among the lower sulfide of a radical, its higher sulfides, and those of its sulfate salts, neutral and basic, that violated the principle. Berzelius clarified his thinking in regard to the neutral salts by investigating degrees of sulfuration and recognizing their correspondence

34 Berzelius, *Versuch*, pp. 99–100.   35 Ibid., pp. 127–135.
36 Ibid., pp. 176–208.   37 As Ostwald points out; Berzelius, *Versuch*, p. 217.

with degrees of oxidation. That clarification is the subject of this section. Berzelius's studies of double salts, which enabled him to explore the violations of the principle, are also treated in a later section.

Berzelius's recognition of a correspondence of oxidation and sulfuration emerged in the first post-"Versuch" collection of analytical studies, published as the bulk of the long paper, here called the "Rechtfertigung."[38] His view was still limited to the correspondence of the lowest oxide with the lowest sulfide, but the discovery of new sulfides later led him to perceive a general correlation, clearly expressed in the next collection of analytical results, "Cause of Proportions."[39] In the "Rechtfertigung," Berzelius exploited a correspondence of lower sulfide with lower oxide as an analytical tool. When the lowest oxide could not be analyzed accurately or reproducibly, he used the sulfide in minimo as a standard against which to determine the progression of oxides. The recognition by his student, Nils Gabriel Sefström (1787–1845) that cinnabar (HgS), hitherto the only known sulfide of mercury, corresponded not to the lower oxide but the higher and his discovery of a lower sulfide (Hg_2S) that corresponded to the lower oxide ended Berzelius's preoccupation with the lowest sulfide and led him thereafter to seek for general correlations of oxidation and sulfuration. Thus, in his *Proportions chimiques,* Berzelius converted this knowledge into the second of three rules that he invoked to determine the formulas of oxides (the first was to examine and compare the oxides of a given radical and the third was the rule of oxides):

> 2nd, one may compare the degrees of sulfuration of bodies with their degrees of oxidation, which [however] do not always correspond. It is known that arsenic, for example, can combine in two proportions with oxygen and with sulfur. The oxygen in these two oxides is as 3 to 5; but the sulfur, in the sulfides, is as 2 to 3; thus the oxide with three atoms of oxygen corresponds perfectly to the higher sulfide, which contains three atoms of sulfur. One may conclude from these considerations that the number of atoms of sulfur and of oxygen in these compounds are 2, 3, and 5.[40]

Furthermore, perception of a general correspondence permitted Berzelius to recognize that to each sulfide of a radical there corresponded a salt having the radical and sulfur in the same proportion as the sulfide.

38 Berzelius, "Versuch, die chemischen Ansichten, welche die systematische Aufstellung der Körper, in meinem Versuch einer Verbesserung der chemischen Nomenclature begründen, zu rechtfertigen," *Journal für Chemie und Physik,* 1812, 6:119–176, 284–322; 1813, 7:43–78.
39 Berzelius, "Cause of Proportions" (in Abbreviations).
40 Berzelius, *Proportions chimiques,* pp. 117–118.

He was then able to link this recognition with the question of maintenance of neutrality upon oxidation of the basic radical (the phenomenon
that had provoked Berzelius's first articulation of the preservation of
combining proportions): If the basic radical is oxidized, neutrality of the
salt is preserved if the salt takes an additional dose of base. The resultant
salt contains the radical and sulfur in the same proportion as the higher
sulfide.[41] His later studies on basic salts (discussed in a later section)
allowed him to describe how the failure of the oxidized radical to take
up more base led to the production of basic salts in violation of the principle.

Water as a Component of Salts

By studying the role of water in salt formation, Berzelius gave further
elaboration to his electrochemical generalization of acidity and basicity,
for he showed that water, *the* neutral substance, could take the role of
either acid or base. In developing techniques for the differentiation of
water of crystallization from water as a saline component, Berzelius proved
that other oxides, not typically acidic or basic, could take over these roles
by displacing acids from salts. Moreover, by studying water of crystallization as a neutral component of salts, Berzelius took the first steps in his
passage from second- to higher-order composition. This work was recounted in the third "Continuation" of the "Versuch."

Davy's denial of the oxide nature of muriatic acid was the occasion for
Berzelius's study of the role of water as base. Impressed by his inability
to decompose muriatic acid into the supposed muriatic radical and oxygen, Davy suggested instead that oxymuriatic acid was an element, chlorine, which, in combination with hydrogen, produced muriatic acid. He
offered this view in opposition to the ideas of Gay-Lussac and Thenard,
who held that muriatic acid, a gas in its ordinary state, contained a quantity of water such that its oxygen equaled that in muriate salts.[42] This
opinion of his French colleagues Berzelius moved to uphold against Davy.[43]

The failure, chronicled by both Davy and the French collaborators, to
prepare dry muriatic acid gas and to decompose muriates in the absence
of water and conversely the ability to do so in the presence of water
constituted proof for Berzelius that water was essential to the existence
of the acid. Muriatic acid thus exemplified a compound in which one
oxide (in this case of the muriatic radical) was incapable of an independent existence, but could exist in combination with other bodies (in this
instance, with water or the bases of muriate salts). The decomposition of

41 "Cause of Proportions," pp. 452–453.
42 Partington, *History,* Vol. IV, pp. 52–54.   43 Berzelius, *Versuch,* pp. 154–158.

muriates in the presence of water in this view amounted to a displacement in which water displaced the base of muriate from combination with the anhydrous acid and combined with the acid to form muriatic acid. Berzelius then suggested that sulfuric and nitric acids also contained oxides incapable of an independent existence and therefore contained basic water (though nitrous and sulfurous acids could exist as anhydrides).

He had thus brought even the mineral acids under his saline view of matter; he also attempted, in the "Versuch," to illustrate the commonplace occurrence of acids capable of existing only in combination with water by analyzing three vegetable acids, thus launching on its long career a view of organic acids that proved dominant until the 1850s.[44]

Berzelius also considered compounds in which water played the role in acid in combination with bases.[45] Some metallic hydrates (now called hydroxides) were already known; the decomposition of the alkalies having revealed that the alkalies and earths were metallic oxides, Berzelius could now establish the existence of a general class of such compounds and incorporate them into his saline theory by treating them as salts of basic metallic oxides with water as acid. Moreover, his quantitative analyses suggested that these compounds were governed by a version of the rule of oxides (that the basic and acidic oxygen were the same).

In the final, summary section of the "Versuch," Berzelius gave forceful expression to the potential of water to assume the role of either acid or base in binary compounds of oxides, but he listed salts containing water of crystallization under the same heading as double salts, compounds containing three or more oxides.[46] This grouping suggests that Berzelius attributed to neutral water of crystallization a role ontologically equivalent to that of acidic and basic saline components; but it also indicates that Berzelius had yet to reach his later, hierarchical model of complex compounds, which regarded double salts as binary compounds of simple binary salts with one another or with single oxides. Instead, he had provisionally settled on a multiple-oxide model of complex compounds that he juxtaposed with the binary model of simple salts.

He formulated a variant of the rule of oxides for salts with water of crystallization to provide stoichiometric control over the difficult experimental task of determining the quantity of water of crystallization. Given a stoichiometric limit on the quantity of crystal water, only approximate experimental results would be necessary to determine the water present. The rule, called here the "law of unitary oxygen," stated that in compounds of several oxides the oxygen of the component containing the

44 Ibid., pp. 160–165.   45 Ibid., pp. 165–177.
46 Ibid., pp. 206–207.

smallest quantity of oxygen is a unit of which the oxygen of each of the other constituents was an integral multiple. Despite what Berzelius explicitly acknowledged to be the provisional status of the rule, he came to adhere to it tenaciously as a device to delimit the number of permissible inorganic compounds (as noted later).[47]

This multiple-oxide view of complex salts persisted throughout Berzelius's studies of basic salts and coexisted with a more nearly dualistic conception. In linking compounds with three oxides together with double salts and in replacing Bergman's designation as "triple salts" of salts having two acids and one base with the coinage "double salt," Berzelius suggested the direction in which he was moving. However, only in his mineralogy, when he had difficulty in bringing still more complex compounds into line with the multiple-oxide model, did Berzelius attain his mature conception of hierarchical dualism.

Basic Salts

Various eighteenth-century chemists had produced acidic and basic salts by combining neutral salts with excess acid or base, respectively, and they held that in some cases the excess exhibited a point of saturation. In *Méthode de nomenclature chimique*, Fourcroy (but not Lavoisier in his *Traité*), accommodated them into the antiphlogistic nomenclature by invoking the modifiers "acidule" and "sursaturé" to the names of the corresponding neutral salts. Berthollet, however, in his famous controversy with Proust, had invoked experiments on basic salts as evidence for indefinite composition.[48] Working in the aftermath of this controversy, Berzelius aimed to exploit his methods to develop a stoichiometry of basic salts.[49]

He began by offering a new, quantitative definition of neutrality, for his own experiments, like several of Richter's, had revealed both salts of weakly basic oxides that exhibited an acidic reactivity incapable of saturation by additional base and salts of weak acids that exhibited a basic reactivity incapable of saturation by additional acid. To accommodate this sort of case, at least in regard to basic salts, he turned the qualitative norm of neutrality into a quantitative one. It was the proportions of the definitely neutral salts that would define neutrality for other salts: An earthy or metallic salt of an acid was neutral if the proportion of basic to acidic oxygen was the same as in an alkaline or alkaline-earth salt of the same acid. The rule of oxides, which had been established with reference

---

47 Ibid., pp. 177–186.
48 Lavoisier et al., *Méthode*, pp. 96–97; Partington, *History*, Vol. III, pp. 648–649.
49 Berzelius, *Versuch*, pp. 186–196; Melhado, *Berzelius*, pp. 224–231.

to indisputably neutral salts of the mineral acids, was now used to define the neutrality of salts formed by relatively weak bases.

Berzelius now examined the ways in which basic salts departed from this quantitative definition of neutrality. To this end, he reanalyzed the salts he had investigated when first articulating the principle of the preservation of combining proportions (for he had understood that basic salts in some manner violated that principle), and he added new analyses of additional examples of basic salts of a variety of acids, including the basic salts of the nitrogen oxyacids, work considered shortly below. His new studies showed that in basic salts the acidic oxygen can be a submultiple as well as a multiple of the basic oxygen. Moreover, he could now complete the argument begun in "Cause of Proportions" about the violations of the preservation of combining proportions presented by basic salts: Further oxidation of the basic radical of a salt destroys the neutrality leading to the production of basic salts.

Berzelius investigated the basic salts of nitric and nitrous acids not only to help establish the stoichiometry of basic salts but also to provide further evidence for the compound nature of nitrogen. Having shown that the neutral nitrates agreed with the rule of oxides only if nitrogen was compound, Berzelius hoped the stoichiometry of the basic nitrates would provide further proof. His study of these substances was flawed by both its predication on Gay-Lussac's value of the composition of nitric acid and Berzelius's difficulties in evaluating the complex events leading to the synthesis of the salts. However, Berzelius did succeed in definitively establishing the existence of the salts of nitrous acid and in preparing a series of both neutral and basic nitrites. The outcomes of these studies, despite their equivocal support for the compound nature of nitrogen, informed Berzelius's reconstruction of Lavoisier's correlation of oxidation and properties.

Double Salts

The case of double salts marked a point of transition between the multiple-oxide model and more nearly dualistic model of salt formation, which portrays salts more complex than neutral salts (that is, more complex than "second-order" compounds) as formed from combination of first-order salts with other compounds of equal or lesser complexity.[50] This is the notion of hierarchical dualism for which Berzelius was so widely noted.[51] In this view, double salts are third-order compounds, formed from combination of two second-order compounds (salts of a

50 Berzelius, *Versuch*, pp. 196–205; Melhado, *Berzelius*, pp. 231–238.
51 The classic statement of the dualistic doctrine is in Berzelius, *Proportions chimiques*, pp. 97–98.

single acid and a single base). Berzelius took this dualistic view of double salts in reflection of their mode of synthesis, mixing solutions of two simple salts, and he regarded these compounds as held together by the mutual affinity of the bases of the constituent simple (second-order) salts. He therefore pressed for a change in their designation from the term triple salts, favored by both Bergman and the French nomenclators dependent on him (because these substances contained three constituents, typically two bases and an acid), to double salts.[52]

However, the model had one deficiency in that it failed to discriminate stoichiometrically between double salts and mere mixtures of simple salts. There was no principle, corresponding to the rule of oxides, that dictated the proportions subsisting between the two constituent simple salts (which Berzelius later described as "second-order" constituents). As shown below, Berzelius's dualistic model amounted to a call for the construction of such rules for higher orders. In their absence, Berzelius relied on another approach, based on the multiple-oxide model. He examined the proportions subsisting among the quantities of oxygen carried by each of the first-order constituents in a double salt and then devised a rule to govern these proportions. By examining three hydrated double salts in the "Versuch," Berzelius devised the "law of unitary oxygen," which stated that the oxygen contents of the various constituents were integral multiples of the oxygen content of the constituent containing the least oxygen. Though he originally presented the rule as only provisionally established, at the conclusion of the "Versuch" Berzelius presented the rule without qualification. Alum, one of the salts Berzelius had used to establish the rule, became his favorite instance, one that proved particularly appealing, because, as both a laboratory chemical and a naturally occurring mineral, alum suggested the applicability of Berzelius's stoichiometry to the mineral realm (a subject treated later).

Oxidation and Properties:
Reconstructing Lavoisier's Correlation

Berzelius's effort to reconstruct the correlation rested on his conviction that the composition of acidic oxides must be linked to that of their salts. The rule of oxides was the bridge between them, and Berzelius's first examples documenting the rule (the sulfur oxyacids and their salts) were at once the source of this conviction and the paradigmatic case of the anticipated connection. The leitmotif of this effort was therefore the sulfur pattern of oxidation and salt formation: Of the four oxides (two incapable of an independent existence), the two acidic ones contained

---

52 Berzelius, *Versuch*, p. 196; Melhado, *Berzelius*, pp. 231–232, text, and nn. 6–7.

oxygen, for a given weight of sulfur in the proportion of 2:3, and the multipliers by which the basic oxygen of their neutral salts was contained in the acids were, respectively, 2 and 3 — that is, the -ous: -ic proportion in the amount of oxygen was 2:3. In neutral sulfites, for example, $PbO \cdot SO_2$, the acid contained twice the basic oxygen; in sulfates, for example $PbO \cdot SO_3$, three times. The oxygen contents of the two oxyacids (sulfuric and sulfurous) were as 3:2, and the acids formed neutral salts so composed that the multiplier by which the basic oxygen of sulfates was contained in the sulfuric acid was to the corresponding multiplier for sulfites also as 3:2. Much of Berzelius's research after the "Versuch" was motivated by the hope that the sulfur pattern might be shared by other radicals.

The appeal of this possibility was increased by Berzelius's knowledge of the neutral salts that he had shown, in the first "Continuation" of the "Versuch," to obey the rule of oxides.[53] In the case of the arsenic compounds, for example, Berzelius analyzed its two metallic acids and the supposedly neutral salts they produced. His analyses suggested to him that, like the two oxyacids of sulfur, the oxyacids of arsenic contained oxygen in the proportion of $1 : 1\frac{1}{2}$ (it is in fact 5:3); that in neutral arsenites (like neutral sulfates but not sulfites), the acid contained three times the oxygen of the base; that in neutral arsenites, the corresponding multiplier was two (as in sulfites but not sulfates); and that, unlike the sulfur oxyacids, arsenites could not be simply oxidized to arsenates with preservation of combining proportions. The similarities between the cases of arsenic and sulfur as well as others involving the oxyacids of phosphorus and the supposed muriatic radical as well as in the oxidation stages of iron and lead induced Berzelius to surmise a connection between the progression of oxidation stages and the composition of salts.

Moreover, he had successfully reduced the sesquialteral proportion of oxygen in the two acidic oxides of sulfur to an integral one by finding the two lower oxides in the muriates of sulfur. These contained, respectively, one-sixth and one-third the oxygen of the sulfuric acid (or one-fourth and one-half that of sulfurous acid; written in modern atomic weights, the four oxides are $SO_{1/2}$, SO, SO_2, and SO_3). Perhaps something similar could be done with other radicals. Again in the case of arsenic, Berzelius compared the proportions in which it combined with other radicals with those shown by, for example, iron and sulfur.[54] Finding anomalies, he supposed that they might be reduced if arsenic could be shown to possess still unknown lower oxidation. Accordingly, he looked for a lower oxide containing one-sixth the oxygen of the highest oxide. Unable to identify it as an ingredient in a compound, Berzelius claimed

53 Berzelius, *Versuch*, pp. 113–117.   54 Ibid., pp. 118–121.

to have discovered it in the black powder into which metallic arsenic degenerates on exposure to air. Berzelius thus aspired to impose a uniformity of stoichiometric phenomena (the sulfur pattern) on a qualitatively diverse series of radicals. The nitrogen oxyacids also proved important in sustaining Berzelius's search for the examples of the sulfur pattern,[55] because new analyses (by Dalton, Davy, and Berzelius himself) of the composition of nitric acid suggested that, if nitrogen is compound, its lowest oxide contains one-sixth the oxygen of its highest, and new studies Berzelius undertook of the neutral and basic salts of the nitrogen oxyacids (which led to a much improved understanding of the basic salts) seemed to provide additional evidence of this kind.

In the lengthy "Essay on the Cause of Chemical Proportions," Berzelius tried to bring the compounds of as many radicals as possible into conformity (in whole or in part) with the sulfur pattern.[56] Though he seemed to have achieved great success in this endeavor, showing, for example, the significant conformity of the arsenic compounds to the sulfur pattern, Berzelius also provided the first evidence (in a study of antimony compounds) for a second pattern, which he later showed characteristic of phosphorus as well and, in a later attempt with arsenic, the compounds of that radical too.[57] The similarity of phosphorus and nitrogen, moreover, and what seemed to Berzelius the unlikely possibility (though one he investigated closely) that phosphorus was similarly compound, led him to accept a second (phosphorus) pattern, which resulted in salts not in conformity with the rule of oxides (but agreeing with a specially formulated variant of it).

Having been compelled to recognize more than one pattern of oxidation and salt formation, Berzelius wrote his *Proportions chimiques* mindful of three categories of radicals: those more or less in conformity with the sulfur pattern, those in conformity with the phosphorus pattern, and those agreeing with neither pattern, but producing salts in agreement with the rule of oxides in its original form. Because radicals of the first and third categories agreed with the rule, Berzelius lumped them together and distinguished them from those of the second category:

> Although this law [the rule of oxides] governs the largest number
> of combinations of compound atoms among oxidized bodies [that
> is, compounds formed by oxides with one another], there are

55 Berzelius, "Neue Untersuchungen über die Natur des Stickstoffs . . . ," *Annalen der Physik*, 1814, 46:131–175; Melhado, *Berzelius*, pp. 241–248, 249, n. 7.

56 Melhado, *Berzelius*, pp. 248–254.

57 Berzelius, "Untersuchung über die Zusammensetzung der Phosphorsäure . . . ," *Annalen der Physik*, 1816, 53:393–446; 1816, 54:31–55; Berzelius, "Nouvelles recherches sur les proportions chimiques," *Annales de chimie*, 1817, 5:174–181, on pp. 179–180; Melhado, *Berzelius*, pp. 254–260.

some exceptions, which however do not occur at random among oxides in general but are confined to certain acids which all share the property that the radical produces two oxides in which the different quantities of oxygen are to one another in the proportion of three to five. These are the compounds of phosphorus, arsenic, and nitrogen. . . . But even for these acids there is a law that governs their combination with other oxides such that the number of atoms of oxygen in the oxide is one or more fifths . . . the number of atoms of oxygen in the acids in -ic and one or two-thirds this same number in the acids in -ous.[58]

With his knowledge of the patterns of oxidation and salt formation, Berzelius completed his evolution away from Lavoisier's classification of oxides. Knowing that the fundamental electrochemical endowment of an oxide rested on that of its radicals, Berzelius nevertheless understood that many radicals shared (at least in part) certain patterns of oxidation and therefore (through the rule of oxides and its variants) salt formation. Degree of oxidation was indicative of acidic character, but within the limits of the electrochemical endowment of the radical. Moreover, by invoking a combination of qualitative and quantitative characters (electrochemical endowment and stoichiometric patterns), Berzelius was in a position to establish the specificity of virtually all compounds encountered in the laboratory, no matter how complex and how unlike they were from the archetypical acids, bases, and salts.

Mineralogy and Organic Chemistry

By the middle of 1813, Berzelius had accomplished the bulk of the research on which the main lines of his chemistry rested, and he had yet to encounter phenomena (indicating the existence of the phosphorus pattern, for example) that would compel him to qualify his views. He had shown to his own satisfaction that the nitrogen compounds conformed to the sulfur pattern and he had succeeded in demonstrating the apparently widespread occurrence of that pattern. Doubtless feeling confident in the value and durability of his achievements, Berzelius attempted deployments of his chemistry in mineralogy and organic chemistry.

Both subjects had been among his earliest interests. From the outset of his career he was closely associated with two figures prominent in Swedish mining and mineralogy, Johan Gottlieb Gahn (1745–1818) and Wilhelm Hisinger.[59] An assistant to Bergman in the 1760s, Gahn had pur-

---

58 Berzelius, *Proportions chimiques*, pp. 35–36.
59 For Gahn, see Olsson, *Kemiens historia*, pp. 151–152; and Partington, *History*, Vol. II, p. 201. For Hisinger, see Sten Lindroth, *Kungl. Svenska Vetenskapsakademiens his-*

sued a distinguished career in the mining industry. Many of his ideas, apparatus, and techniques, about which he himself had published nothing, entered the literature of chemistry through his association with Berzelius; and during the years 1813–1816, when Berzelius began to pursue mineralogy in earnest, Gahn served as Berzelius's host in the summers at the mines at Falun. Hisinger, a wealthy mining industrialist with well-developed interests in mineralogy and geology, collaborated in one of Berzelius's earliest published studies, describing the discovery of cerium. Thereafter, Berzelius published several papers annually on mineralogical topics. Also during the years prior to the "Versuch," Berzelius devoted much attention to chemical investigation of organic matter, especially animal products. A book recounting his research in animal chemistry appeared in Stockholm in 1806–1808, and a collection of numerous analyses of animal substances that he published in England while visiting there made his work widely known.[60] His lecture on the state of animal chemistry, given upon his relinquishing the presidency of the Royal Swedish Academy of Sciences in 1810, also became widely known through translations of the published Swedish version.[61] These interests in mineralogy and organic chemistry had become eclipsed by Berzelius's involvement in the research reported in the "Versuch" and its direct successors, but during the winter of 1813, when his work reached a temporary culmination in the "Cause of Proportions," Berzelius returned wholeheartedly to these subjects.[62]

His aim was to invoke in these fields the same combination of qualitative and quantitative characters that he had used to establish the specificity of inorganic laboratory chemicals. Berzelius's mineralogy exhibits this approach very clearly. By recognizing the acidic role of silica, Berzelius could assimilate the numerous siliceous minerals to the category of salts by portraying them as complex silicates. If he could show the conformity of their proportions to the laws of stoichiometry, Berzelius could use his chemistry to specify every mineral known. The stoichiometric approach put Berzelius into conflict with Wernerian mineralogists. Allying himself with René Haüy, Berzelius claimed that, in mineralogy, stoichiometry defined species unequivocally without reference to the doctrine of transitions. His case was strong but not immune to his Wernerian

toria 1739–1818, 2 vols. (Stockholm, 1967), Vol. II, pp. 390–393; and C. Heijenskjöld, "Wilhelm Hisinger," *Med Hammare och Fackla*, 1933–1934, 5:56–72.

60 Berzelius, *Föreläsningar i djurkemien*, 2 vols. (Stockholm, 1806–1808); Berzelius, "On the Composition of Animal Fluids," *Annals of Philosophy*, 1813, 2:19–26, 195–208, 377–387, 415–425; cf. Rocke, Chapter 5 in this volume.

61 Berzelius, *A View of the Progress and Present State of Animal Chemistry*, tr. Gustavus Brunmark (London, 1813; reissued, 1818).

62 Melhado, *Berzelius*, p. 260, n. 7.

opponents, who responded much as Berthollet had responded to Haüy. Though Haüy had ceded some ground to his critics, having recourse to morphology in the determination of species, Berzelius hoped to meet his critics with new articulations of his chemistry, including his mature doctrine of electrochemical dualism.

Berzelius hoped to proceed similarly in organic chemistry, but his successes there, if no less important, were more limited. His two chief innovations (up to the *Proportions chimiques*) were, first, to argue that the fundamental level in organic chemistry must be the same as in inorganic chemistry, that is, the specific compound, composed in definite proportions (as opposed to the traditional and unstable collections of extracts traditionally obtained in organic analysis); and, second, to identify the obstacles to attainment of the specific level and to account for them: Because organic compounds were subject to fewer stoichiometric restrictions than inorganic compounds, definite proportions did not prohibit transitions. Organic compounds could therefore occur as generic mixtures (which Berzelius identified with the products of traditional extractive analysis), but access to species demanded resolution of the genus, and Berzelius saw no way to accomplish that resolution. He therefore conceded that organic chemistry, unlike mineralogy, would have to remain part of natural history.

Organic Chemistry

Berzelius's organic chemistry therefore can be regarded as an effort to identify the extent to which stoichiometry governs the composition of organic matter. In his organic analyses antedating the "Versuch," Berzelius had shown consummate skill in the use of extractive methods; but his own sensibilities as an observer made him doubt that the products obtained by extractive analysis amounted to anything fundamental (e.g., constituting something like Fourcroy's "immediate materials").[63] After the "Versuch," Berzelius understood that the task facing organic analysis was to investigate the extent to which stoichiometric law governed the composition of organic matter, for he could see no other foundation for discrimination of kinds. This effort contained both theoretical and experimental sides. The former consisted of repeated efforts to assess on the level of principle the extent to which organic matter is subject to the law of unitary oxygen (and its near-translation into atomic terms, the

---

63 See, for example, Berzelius, "Versuche über die Mischung des isländischen Mooses und seine Anwendung als Närungsmittel, *Journal für Chemie und Physik,* 1813, 7:317–352.

law of the unitary atom).[64] In 1813, responding to criticism leveled at his stoichiometry by Dalton, and repeatedly thereafter, Berzelius upheld his unitary laws by arguing that without them "all ideas of determinate proportions would disappear."[65] The laws did not merely allow discrimination of species; by limiting the proliferation of species, they rendered the determination of species possible.

His thinking having undergone subtle changes from 1813, Berzelius eventually concluded that the locus of difference between organic and inorganic chemistry lay in the first order.[66] In a manner reminiscent of Lavoisier, Berzelius argued that organic compounds consist of oxides of complex radicals (formed not by binary, but stoichiometrically unlimited ternary and higher combinations of the elements), which then could enter into (stoichiometrically delimited) second- and higher-order compounds. The experimental foundation for Berzelius's views lay in his demonstration that salts of a few easily isolable vegetable acids exhibited the same sort of second-order stoichiometry as analogous salts of inorganic acids.[67] The higher-order composition being the same, the locus of difference must be on the first order.

Organic matter therefore exhibited transitions even though it consisted of compounds in definite proportions. Small changes in composition produced minuscule changes in properties. As a result, traditional extractive analysts had produced generic mixtures, which, on Fourcroy's model, they had regarded as the ingredients of plants. Berzelius, however, warned that quantitative analysis of the products of extraction was useless unless generic mixtures could be resolved into individual stoichiometric compounds. At the end of the second decade of the century, the prospects for that kind of resolution did not look good. Organic chemistry, it seemed, would have to remain with natural history.

Mineralogy

By contrast, mineralogy seemed amenable to stoichiometry. As noted earlier, with cases like that of alum, Berzelius had already shown that his

64 For the atomic version of the unitary law, see Melhado, *Berzelius*, pp. 270–274, 308, 308n.
65 Ibid., pp. 276–277, 277, n. 4; the quotation is from Berzelius, "An Address to Those Chemists Who Wish to Examine the Laws of Chemical Proportions, and the Theory of Chemistry in General," *Annals of Philosophy*, 1815, 5:122–131, on p. 124.
66 Melhado, *Berzelius*, pp. 283–293.
67 Berzelius, *Versuch*, pp. 160–165; Berzelius, "Experiments to Determine the Definite Proportions in Which the Elements of Organic Nature Are Combined," *Annals of Philosophy* (1814), pp. 4:323–331, 401–409; 1815, 5:93–101, 174–184, 260–275; Berzelius, *Proportions chimiques*, pp. 40–47; Berzelius, *Lehrbuch der Chemie*, tr. Friedrich Wöhler, 4 vols. in 8 (Dresden, 1825–1831), Vol. III:i, pp. 135–175.

chemistry could cross the line from the laboratory to the mineral realm. At least two additional factors also encouraged Berzelius to pursue mineralogy from a stoichiometric standpoint. First, the decomposition of the alkalies and earths, achieved by Berzelius and Davy during the first decade of the nineteenth century, allowed him to eliminate the traditional mineralogical class of earths and classify the large number of earthy minerals in the same way as compounds of metallic oxides were classified.[68] Second, his generalization of the concepts of acidity and basicity allowed him to regard the weakly reactive oxides that entered into the composition of minerals as saline constituents.[69] These included the oxides of titanium, tantalum, and silicon (silica). This last was particularly important for mineralogy. Berzelius (simultaneously with Smithson) proposed that silica, regarded in the eighteenth century as an alkaline earth, was a weakly reactive acid, a view permitting him to portray a vast number of minerals as silicates and confer on them systematic names according to the principles of the new nomenclature. If most minerals were salts of weak acids and bases, their composition should conform to stoichiometric law, just as did those (like alum) that were also laboratory chemicals. Both the qualitative and quantitative dimensions of Berzelius's view of salts seemed applicable to even complex minerals. Berzelius was therefore emboldened to subordinate mineralogy to chemistry.

He made this attempt in two small treatises published in 1814 and 1815.[70] The former he offered as a set of illustrated principles; the latter, as a more fully articulated mineral system that aimed, in part, to counter Wernerian criticism elicited by the former. In both, he proceeded from a strict, stoichiometric definition of mineral species.[71] In the 1814 treatise, he developed the argument that, from his knowledge of weakly reactive oxides and from the example of alum (at once a mineral and a laboratory chemical, which he had successfully portrayed in terms of the multiple-

68 Berzelius, *Versuch*, pp. 119–120, 120, n; Berzelius, "Versuch eines rein chemischen Mineralsystems," *Journal für Chemie und Physik*, 1815, *15*:301–363, 419–451; 1818, 22:274–302 (henceforth cited as "1815 Mineralogy"), on pp. 361–362; Melhado, *Berzelius*, pp. 296–300.

69 Berzelius, "Versuch, durch Anwendung der electrisch-chemischen Theorie und der chemischen Proportion-Lehre ein rein wissenschaftliches System der Mineralogie zu begründen," *Journal für Chemie und Physik*, 1814, *11*:193–233; 1814, *12*:17–62; 1815, *15*:277–300 (henceforth cited as "1814 Mineralogy"), on 198–202, 212; Melhado, *Berzelius*, pp. 298–301.

70 The two treatises, "1814 Mineralogy" and "1815 Mineralogy" just cited. The English translation of the former was cited earlier, near the end of the discussion of eighteenth-century mineralogy; for further bibliographic details, see Melhado, *Berzelius*, pp. 293n–295n. For mineralogy as a branch of chemistry, see "1814 Mineralogy," pp. 194–198; and cf. Melhado, "Mineralogy and the Autonomy of Chemistry."

71 Berzelius, "1814 Mineralogy," p. 231; "1815 Mineralogy," p. 422.

oxide model), mineralogy could be regarded as merely a branch of chemistry. Advances in the knowledge of weakly reactive oxides and in methods of analysis showed that complex substances were fully amenable to designation according to the principles of nomenclature already successfully applied to simpler salts; and he offered such a nomenclature for complex silicates. However, in the same treatise, Berzelius began to get some presentiment of the special problems that made mineralogy resistant to stoichiometry, and Wernerian criticism of his mineralogy turned his presentiment into full recognition. In the second treatise, he modified his position, offering instead of the multiple-oxide model the fully dualistic view of chemical combination that came to characterize his chemistry. Berzelius's concern for the specific and the particular demands of mineralogy thus combined to occasion a significant articulation of chemical theory.

The Wernerian J. F. L. Hausmann (1782–1859) argued that Berzelius's system was "wholly unmineralogical" because in discriminating species it did not take into account affinities exhibited in external characters.[72] An assessment of the influence of composition on external characters was in order, not complete reliance on composition. By this procedure, "it is possible to regard as mineralogically nonessential certain constituents that from a purely chemical standpoint are no less essential than others." With a purely stoichiometric perspective, the "chemist . . . will often separate certain species that appear to the mineralogist as only varieties of one and the same substance." What Berzelius took to be significant differences in composition produced insufficient alteration in external characters to justify separation of species. "No scientific mineralogist" would follow Berzelius in drawing specific distinctions between samples with great external affinity.[73] This easy recapitulation of the standard Wernerian criticism of chemical mineralogy shows that, for all its thorough articulation, Berzelius's elaborate stoichiometry was still incapable of providing mineralogy a persuasive discrimination of compounds from mixtures among the products of the mineral realm.

Berzelius responded to Hausmann in the treatise of 1815 with a series of "Betrachtungen über die Constitution der Mineralien" [Considerations on the constitution of minerals].[74] There, he recognized the inadequacy of the multiple-saline model of the earlier treatise and offered instead his mature expression of the dualistic doctrine. He began by introducing the distinction between fine mixture and combination, and

---

72 J. F. L. Hausmann [Review of the Swedish edition of the "1814 Mineralogy"], *Göttingische gelehrte Anzeigen, 1814,ii*:1089–[1101].

73 For more about the Wernerian response to stoichiometry, see Melhado, "Mitscherlich's Discovery."

74 Berzelius, "1815 Mineralogy," pp. 323–336.

he asserted that the discrimination of compounds from mixtures amounted to the problem of partitioning the oxides revealed by analysis either among several different compounds (indicating mixture) or among the lower-order ingredients of a higher-order compound (indicating a pure substance). He held that the task would be simplified if it were known whether there was a limit on the number of second-order constituents that could enter into a compound, and he suggested that an answer might be obtained by confining scrutiny to members of a single crystallographic species. Finally, he provided his first full statement of the dualistic doctrine. In inorganic nature, he observed, elements combine into binary entities according to their electrochemical opposition, and the products of such unions could themselves enter into binary combinations. He described salts with water of crystallization as compounds of the anhydrous salts with water. As for alum, formerly his chief example of the multiple-oxide model (a combination of four oxides), he now regarded it as a compound of anhydrous alum and water, and he described the anhydrous alum in turn as a compound of two simpler salts. In the case of a more complex mineral, he continued to invoke the unitary law, which continued to perform its function as the delimiter of speciation, but, like the rule of oxides for second-order compounds (simple salts), he subordinated it to a fully dualistic view: At each level in the dualistic hierarchy, partitioning must take place according to a unitary law. As the passage suggests, Berzelius did not suppose he had thus solved the problem of discriminating compounds from mixtures; more than one life's work was needed and the task was no doubt daunting. However, in this articulation of his thought he found what he hoped would secure the applicability of stoichiometry to mineralogy and serve as the chemical foundation for identifying essential ingredients and exhibiting the specificity of compounds.

Conclusion

Berzelius's work must be understood by comparing it to its predecessors. In a variety of ways, Berzelius was merely carrying forward the eighteenth-century interest, evident in a variety of contexts, in discriminating kinds of matter and was doing so in the intellectual context provided by the new chemistry of Lavoisier and the invention of the Voltaic pile. On the other hand, his achievements rested on the heavier stress, perhaps deriving from the Swedish mineralogical traditions, that he placed on the importance of species of matter (as opposed to generic groups). That stress manifested itself in important novelties that distinguished Berzelius from his French predecessors, especially his relocating the seat of properties from the supposed generic ingredient to the radical, his elevation of basicity to an equal status with acidity, his electrochemical generalization of the concepts of acidity and basicity, his separation of the role of the radicals in the determination of properties from that of degree of oxidation, his reconstructed correlation of oxidation and properties, and his devising the mature dualistic doctrine to accommodate mineralogy. Moreover, Berzelius pursued these qualitative innovations in close conjunction with his own stoichiometry. Unlike Dalton, who articulated a few of the principles of stoichiometry, Berzelius strove to elaborate a detailed stoichiometry, epitomized in the various laws he articulated on the basis of comprehensive empirical investigation and applicable to various classes of compounds. By characterizing substances both qualitatively (as either saline constituents, i.e., electrochemically acidic or basic oxides or sulfides or salts) and quantitatively (as conforming to stoichiometric law), Berzelius could establish the specificity of virtually every inorganic chemical known, including a large number of minerals. He was also in a position to lay the foundations for organic chemistry by showing that past analytical practice rested on inadequate foundations and aiming to establish as a new foundation the stoichiometrically defined species of organic matter.

This image of Berzelius's system was attained by looking at its construction rather than by dissecting the fully articulated system as nineteenth-century authors (and those inspired by them) tried to do. The traditional authors sought in this way to show that Berzelius's system aimed to advance goals set by his predecessors, especially Lavoisier and Dalton, and thus laid it open to attack by a new generation. However, because these authors began with the developed system, they paradoxically missed the main concern that Berzelius shared most fully with his forebears (the desire to discriminate kinds), they failed to see the ways in which Berzelius's system both accommodated and advanced that goal, and they did not recognize how prepared Berzelius was, having anchored himself to a traditional goal, to reach it by departing from tradition where necessary. His work was thus innovative as well as traditional, and the rigorous stoichiometric foundation he provided chemistry constituted one of its enduring foundations. If the next generation proved willing to depart from Berzelius (as he did from his own predecessors), that is scarcely surprising; but Berzelius's successors arrived at their point of departure only after fully exploiting in the field of organic chemistry the legacy that he had provided them. Moreover, what they left behind was not the very real gains thus achieved, but the conviction that Berzelius's chemistry could adequately accommodate the diverse phenomena of organic combination. In constructing his chemistry, Berzelius was naturally enough not prescient, but he was surely more innovative and his work more enduring than the next generation could recognize.

7

Berzelius as Godfather of Isomorphism

HANS-WERNER SCHÜTT

After the discovery of chemical isomorphism by Eilhard Mitscherlich (1794–1863) in 1818, Berzelius served as its spiritual godfather. It was he who introduced the "newly born" phenomenon into the scientific world. He was in an excellent position to do so, for he not only liked the "baby," but he could also offer valuable help, and at least in Germany he was considered to be the most important "arbiter chemiae" of his time. Berzelius's influence also bore on cultural policy in Germany. So Mitscherlich wrote on his reputation in Prussia: "He was [himself] previously called here [to be Klaproth's successor at Berlin] with the offer of an enormous salary; the Minister places absolute confidence in him and gave him complete liberty to propose [candidates] to succeed to the foremost chemical position in the state, that is, Klaproth's."[1]

However, outside the chemistry circles of Berlin, Mitscherlich was virtually unknown. He had studied oriental languages and history and had begun to study medicine, too, for he hoped thus to get a passage to Persia as a ship's doctor. In 1814 he had written a doctoral thesis on Persian history, but in the course of time he had found science more appealing than history and so he moved from Göttingen to Berlin to complete his studies in medicine and sciences. There he was particularly supported by the botanist and chemist Heinrich Friedrich Link (1767–1851), who also permitted him to use his private laboratory for chemical studies. In this laboratory he began with his research on phosphorus and arsenic salts that resulted "immediately after Christmas" in the discovery of the isomorphism of these salts.[2]

When Berzelius visited Berlin in September 1819, Mitscherlich, whose

1 Letter from Mitscherlich to his father, 17 September 1819, in *Gesammelte Schriften von Eilhard Mitscherlich*, ed. Alexander Mitscherlich (Berlin: Ernst Siegfried Mittler und Sohn, 1896), pp. 43–44; this, and all other translations, unless otherwise noted, are mine.
2 Hans-Werner Schütt, *Die Entdeckung des Isomorphismus. Eine Fallstudie zur Geschichte der Mineralogie und der Chemie* (Hildesheim: Gerstenberg, 1984), pp. 129–137.

ideas on the interdependence of crystal form and chemical composition were already known among Berlin scientists, was introduced to him. Berzelius was especially interested in the young chemist's work, as Mitscherlich had taken up a topic that was part of Berzelius's own scientific program. In fact, Mitscherlich originally had no other aim than to check if certain results of analyses obtained by Berzelius were right. In 1818 Berzelius had just corrected previous analytic results that had suggested the composition AsO_6 for arsenic acid and PO_2 for phosphoric acid. Now Berzelius found the composition AsO_5 (today, As_2O_5) for arsenic acid and the composition PO_5 (today, P_2O_5) for phosphoric acid. The new results meant that the ratio of oxygen in arsenic acid and in phosphoric acid compared with the oxygen of arsenous and of phosphorous acid (AsO_3 and PO_3; today, As_2O_3 and P_2O_3) had to be assumed to be 5:3. This contradicted the "sulfur pattern," that is, Berzelius's own assumption that the ratio of oxygen in the base of the "-ic" and "-ous" salts should be 3:2. Furthermore, the ratio of oxygen of both neutral salts seemed to violate his "rule of oxides." As Berzelius mistakenly assumed that the base of these salts contained two proportions of oxygen, the ratio of oxygen in the base and the acid of the salts seemed to be 2:5 and 2:3, which means that the oxygen in each of the acidic components was not an integral multiple of the oxygen in the basic constituents. The problem raised here concerned both the arsenic and the phosphorus compounds, and therefore both groups moved into the focus of research together.[3]

Mitscherlich's research not only proved that the ratio of oxygen of the two oxides of arsenic and phosphorus was correct, it also showed that the respective salts of arsenic and phosphorus had the same crystal form. After having compared several other pairs of salts, too, Mitscherlich risked putting forward the inductive generalization that "compounds in which the number of atoms is the same, also have the same crystallization."[4] Later, in 1821, he modified his views, stating that "the same number of atoms combined in the same fashion produces the same crystal form, and that the same crystal form is independent of the chemical nature of the atoms and is determined only by the number and the relative position of the atoms."[5]

3 Evan M. Melhado, "Mitscherlich's Discovery of Isomorphism," *Historical Studies in the Physical Sciences*, 1980, *11*:87–123, on pp. 111–113.

4 Eilhard Mitscherlich, "Über die Krystallisation der Salze, in denen das Metall der Basis mit zwei Proportionen Sauerstoff verbunden ist," *Abhandlungen der Kgl. Akademie der Wissenschaften in Berlin*, 1818–1819:427–437, on p. 427.

5 Eilhard Mitscherlich, "Sur la relation qui existe entre la forme cristalline et les proportions chimiques. IIme mémoire sur les arséniates et les phosphates," *Annales de chimie*, 1821, *19*:350–419, on p. 419.

Both assertions amounted to a direct attack on the mineralogical doctrines conceived by René Just Haüy (1743–1822). These doctrines, which held an almost paradigmatic and exclusive position in mineralogy, were the basis of Haüy's taxonomy. They in turn rested on the assumption that a specific chemical compound can have only one "primary crystal form" and that different chemical compounds cannot share the same crystal form, thus excluding both isomorphism (the condition in which different solid chemical compounds exist in one common crystal form) and polymorphism (the condition in which a solid chemical compound exists in more than one crystal form).[6]

This is not the place to describe Haüy's and his disciples' controversy with Mitscherlich. It should only be stressed that Berzelius without hesitation sided with Mitscherlich against his old and respected colleague Haüy (to whom in 1819 he dedicated the French edition of his *System för Mineralogien . . .*).[7] Obviously, Berzelius at once realized the impact the new discovery had on chemistry as well as on mineralogy, and he regarded Mitscherlich's findings to be "the most important ones since the establishment of the theory of the chemical proportions."[8]

Those who know the papers that Berzelius had previously written on theories of matter, on chemical proportions, and on chemical affinity will not be astonished that it was he who immediately recognized the discovery of isomorphism by a young chemist who, before that time, had never published anything on chemistry. Berzelius's theory of matter does not exclude the possibility of an isomorphism of the crystal forms of various chemical compounds, and furthermore Mitscherlich's discovery promised to contribute to the solution of important problems within the framework of Berzelius's concepts. In chemistry, for example, the determination of atomic weights was especially difficult in cases where measuring the vapor density was not possible; very often one had to depend on conjectures. Now in many cases atomic weights could be determined via isomorphic compounds without many chemical complications. As an example, the element selenium, discovered by Berzelius in 1818, when forming sodium selenate crystallizes isomorphically with the correspond-

6 The fact that certain highly regular "formes limites" are shared by different minerals does not contradict Haüy's basic assumption, as those minerals do not form mixed crystals in all proportions (solid solutions); Hans-Werner Schütt, "René Just Haüy und die Entwicklung der Kristallographie zu einem konstitutiven Teilgebiet der Mineralogie," in *Disciplinae novae. Zur Entwicklung neuer Denk- und Arbeitsrichtungen in der Naturwissenschaft*, et. Christoph Scriba (Göttingen: Vandenhoek and Ruprecht, 1979), pp. 75–89.

7 The dedication of the French translation of the "System" reads: "À Monsieur René-Just Haüy, dont le génie a élevé la minéralogie au rang des sciences." Berzelius, *Nouveau Systéme des Minéralogie* (Paris, 1819).

8 Berzelius, *JB*, 1822, 1:62–74, on p. 74.

ing sodium sulfate. Thus, as the atomic weight of sulfur was known, the atomic weight of selenium could easily be determined, for if sodium sulfate contains one unit of sulfur, isomorphic selenium sulfate contains one unit of selenium. It should be added, though, that this reasoning takes for granted that the relative weight of the sulfur unit in sodium sulfate stands for an *atomic* weight and that one unit of selenium really represents one chemically ultimate particle of selenium. Actually, the quantity of oxygen that is bound by an element x of the quantity y is no criterion for the number of atoms in the oxide. In other words, an inexact differentiation between atomic weights and equivalent weights caused difficulties. In 1828, however, Berzelius showed new ways by which, at least for heavy elements, reliable atomic weights could be obtained. He relied on analogies from compounds of known stoichiometric composition, on Mitscherlich's law (actually a rule), and on the law of Dulong and Petit.[9]

In mineralogy the old problem of subspecies and mixed crystals found a more plausible solution, at least in the eyes of a chemist, though Berzelius was forced to give up the traditional definition of a mineralogical species. (The ensuing controversy on the new and less rigid concept of species is discussed later.) The obvious advantages Berzelius saw in the new discovery, especially for stoichiometric chemistry, probably convinced him immediately of its value.

Without delay he tried to propagate "Mitscherlich's law" by wielding his influence on four different levels. He invited Mitscherlich to Stockholm to give him thorough training in analytical chemistry (December 1819–November 1821); he wrote numerous letters trying to convince his French colleagues of the significance of the new findings; he used his *Annual Reports* (*Jahresbericht*) as a medium to support Mitscherlich in public; and finally, he tried to prove the scientific fruitfulness of isomorphism by supporting studies by his disciples. Here particularly the work of Heinrich Rose (1795–1865), Nils Gustav Nordensköld (1792–1866), and Count Hans Gabriel Trolle-Wachtmeister (1782–1871) on pyroxenes and of Per Adolph von Bonsdorff (1791–1839) on amphiboles should be mentioned.

In his first *Jahresbericht* (1822) Berzelius reviewed these studies as part of a long report on Mitscherlich's discoveries.[10] The report culminated in the sentence: "Mitscherlich's discovery profoundly affects the theory of chemistry, and at this moment it is still impossible to foresee which

9 Berzelius, "Atomgewicht der einfachen Körper," ibid., 1828, 7:67–69; Alan J. Rocke, *Chemical Atomism in the Nineteenth Century from Dalton to Cannizzaro* (Columbus: Ohio State University Press, 1984).

10 Berzelius, *JB*, 1822, 1:62–74; Schütt, *Die Entdeckung des Isomorphismus*, pp. 170, 238–284.

consequences may follow in the future."[11] In the *Jahresbericht* of 1823 Berzelius reported the discovery of the dimorphism of sodium dihydrogen phosphate (NaH_2PO_4); in 1824 he dealt with Rose's studies on the possible isomorphism of analcime and leucite and on the role of crystal water in the formation of isomorphic compounds; in 1825 he described the dimorphism of sulfur. In this report on allotropic modifications of sulfur, Berzelius also published a private note by Mitscherlich on the irregular thermal expansion of noncubic crystals. One year later, in 1826, Berzelius reported on Mitscherlich's further studies showing that crystal angles could be affected by the mode of chemical preparation, for instance, a surplus of acid during the crystallization.[12]

Berzelius also used his vast correspondence to influence his colleagues in Mitscherlich's favor. Two series of letters are of particular interest, those with Haüy and with Pierre Louis Dulong (1795–1838). Berzelius knew that it would be hard to convince Haüy of Mitscherlich's discoveries. All the same he tried to do so, albeit cautiously. So he began a letter in April 1820 to Haüy with the introductory assurance that he was happy to have the right to count himself among Haüy's students, and then continued to give a detailed account of the work of his own disciple Mitscherlich, concluding: "These discoveries shed a bright light on mineralogy and will give the key to the explanation of the discrepancies between chemical analysis and crystallography, because in a substance where the latter has determined the crystal form, the number of atoms and their kind of combinations must be absolutely identical."[13]

Haüy answered by arguing that every mineralogical species has its own specific integrant molecule and therefore its own specific primary crystal form, that the exceptions to the rule in the case of cubic crystals ("formes limites") are well defined, that Mitscherlich has confused those exceptions with truly isomorphic crystals, and that some crystals that Mitscherlich thought isomorphic clearly show differences in the inclination of their faces (up to about 3° in the case of baryte and coelestine). In the autumn of 1820 Berzelius wrote in response to Haüy's letter that he had drawn all objections to the attention of Mitscherlich, who consequently doubled his efforts to present the most precise results. Berzelius then reported on analyses of pyroxene.

Apart from pressing the specific points, he used all his prestige to back his disciples' work, while emphasizing at the same time that the fame was

11 Berzelius, *JB*, 1822, 1:72.
12 Berzelius, *JB*, 1823, 2:41–43; 1824, 3:151–152; 1825, 4:71–73; 1826, 5:180–185.
13 Berzelius to Haüy, April 1820, in Alfred Lacroix, "Tentative de recherches de la correspondance de l'Abbé Haüy," *Bulletin de la Société française de minéralogie*, 1944, 67:193–226, on pp. 205–208 (207).

due not to him but to them: "When saying *we*, I must add that my contribution consists only in guiding the analytical methods, while the main discovery is due to Mr. Mitscherlich, and the idea of verifying it by such a long series of analyses is due to the young chemists who are occupied with these analyses."[14] In both of his following letters Berzelius did not refer directly to isomorphism. In the first of these letters, however, he mentioned a promised consignment of pyroxenes that he had just dispatched. In October 1821, Berzelius received a response from Paris. To explain the different results in the analyses of pyroxene samples, Haüy once more stressed that in his opinion foreign substances are embedded among the integrant molecules. And he categorically rejected Mitscherlich's ideas, appealing to the scientific community as a whole to give testimony to the justification of his own judgment. These disappointing sentences were the last that Haüy wrote to Stockholm before his death.

While Berzelius and Haüy treated each other with respect and reservation, the relation between Berzelius and Dulong was truly amicable. Thus it is not astonishing that Berzelius already in early November 1819 wrote his friend of the new discovery, remarking that it would fit very well into Dulong's own conception:

> You see that this kind of research will add to your work and verify our hypothetical ideas on the number of the elementary atoms which are contained in a compound and on the modes of their combination, while at the same time a part of the favorite ideas of the famous Haüy is refuted by it. The results of Mitscherlich's research will stir up a lot of confusion in mineralogy at the beginning, but one will see more clearly that chemical analysis alone will be able to determine what every mineral is.[15]

Dulong answered evasively: "I now know Mr. Mitscherlich's memoir. Reading it has increased further the interest which your announcement has inspired in me. Mr. Haüy has attacked him a little too superficially. . . . It would be irritating if one finds only a similarity in the forms of samples having different chemical nature but the same proportions."[16] But Berzelius did not want to drop the subject, and so he briefly mentioned Mitscherlich once again in the postscript of a letter dated December 1820: "Mitscherlich is still engaged in his idea, which will be rectified in the light of the criticism it receives, but seems to me too well founded to be refuted."[17] Berzelius did not get an answer to this remark and ap-

14 Berzelius to Haüy, Autumn 1820, ibid., p. 213.
15 Berzelius to Dulong, 5 November 1819, *Bref*, Vol. II, pp. 9–11, especially on pp. 10–11.
16 Dulong to Berzelius, 2 October 1820, ibid., pp. 16–21, on p. 18.
17 Berzelius to Dulong, 18 December 1820, ibid., pp. 21–24, on p. 24.

parently did not expect one. In his correspondence with Dulong he did not mention Mitscherlich again. I do not know if only his friend's lack of enthusiasm caused his reservation, but I think it's more probable that Berzelius soon found it unnecessary to use his correspondence to plead for his disciple. He knew that on his own Mitscherlich would relatively soon win the recognition he merited. In fact a memoir Mitscherlich had published as early as 1822–1823 on the allotropic modifications of sulfur immediately found the approval of the influential Joseph Louis Gay-Lussac (1778–1850).[18]

The story of Berzelius's fight for the recognition of chemical isomorphism has a sequel that should be mentioned, as it shows the difficulties especially mineralogists had to cope with when trying to accept and integrate the new notion. A main problem consisted in the redefinition of the mineralogical species as conceived by Haüy, for the discovery had complicated the relation between the crystallographic form and the chemical content of crystals. In the eyes of Haüy a specific mineral consisted of microcrystals, the polyhedric "molécules intégrantes." These integrant molecules were the smallest units containing all chemical components of the mineral in definite proportions. The macroscopic form of a crystal was determined jointly by the shape of the integrant molecules and the mode of their aggregation. Because the shape of the integrant molecule is determined by its chemical composition in such a way that a specific chemical compound produces a specific and singular polyhedron, the form of the macrocrystal also belongs to but one chemical composition. This means that for Haüy the mineralogical species is defined by a strict correlation of chemical composition and crystal form. Both genuine isomorphism and polymorphism could not be accepted by Haüy. If chemical analyses show varying composition of samples having the same crystal form, Haüy claimed that these differences result from accidental immixtures (*principes accidentels*), which may impose their specific crystal form on substances that normally crystallize differently.[19]

After the discovery of isomorphism these seemingly foreign substances could very often be interpreted as isomorphic and thus substantial components of the crystal. And as several chemical compounds could form crystals of the same shape, the link between chemical composition and crystal form could no longer be considered to be simple, for the crystal shape is bound to the stoichiometric constitution of a compound and not

18 Letter from Gay-Lussac to Mitscherlich, 6 July 1822, Mitscherlich, *Gesammelte Schriften*, p. 194.

19 Even before the discovery of isomorphism the hypothesis of accidental immixtures, which was a natural consequence of Haüy's rigid definition of species, had been a subject of dispute among the followers of Werner, Berthollet, and Haüy; Melhado, "Mitscherlich's Discovery," pp. 88–98.

directly to the compound itself. This means that compounds, which have the same stoichiometric constitution (including molecules of crystal water) and crystallize isomorphically, may intermingle in all proportions. It also means that a classification following chemical criteria and a classification according to crystallographic viewpoints no longer yield the same result. If one classifies crystallographically, polymorphic substances have to be classified as being of different species, whereas isomorphic mixtures having the same form, such as pyroxenes, continue to belong to one species. If, on the other hand, one classifies chemically, polymorphic substances like sulfur still belong to one species, whereas pyroxenes "dissolve" into a series of species according to the ratio of their mixtures. As a chemist, Berzelius chose the second alternative. He was thus forced to augment considerably the number of species or, if he considered only the pure compounds to represent a species, to multiply indefinitely the number of subspecies. Berzelius knew fully well that the species concept had thus lost its significance as the basis of mineralogical taxonomy. In 1825 he implied that it was a free decision to call an isomorphic mixture a species or a subspecies.[20]

For this conclusion and for his attempt to establish a new classification system, Berzelius was heatedly criticized by Haüy's former disciple François Sulpice Beudant (1787–1850).[21] In 1817 Beudant had used Haüy's hypothesis of *principes accidentels* to explain his own findings that less than 10 percent iron sulfate may determine the crystal form of a mixture with 90 percent copper sulfate.[22] But now, after 1823, even he accepted isomorphism at least as an "analogie très grande" of chemically different compounds.[23] The "form-giving capacity" of iron sulfate could now be explained differently: Copper sulfate is liable to polymorphic crystallization, and one of its crystal forms crystallizes isomorphically with iron sulfate.

After 1825 not only Haüy's faithful disciple Beudant but virtually all mineralogists and chemists took the phenomena of isomorphism and

20 "Wir haben also hier Gattungen, Arten und Varietäten, oder, wenn man mit dem Namen Gattungen nicht einverstanden ist, haben wir Arten, Unterarten und Varietäten." Berzelius, "Über jene im chemikalischen Mineralsystem nöthigen Veränderungen, welche sich daraus ergeben, daß isomorphe Körper einander in unbestimmten Verhältnissen zu ersetzen vermögen," *Archiv für die Gesammte Naturlehre*, 1825, 4:1–46, on p. 7.
21 François Sulpice Beudant, "Sur la classification des substances minérales," *Annales de chimie*, 1826, 31:181–205, 225–243; Hans-Werner Schütt, "Beudant, Berzelius und die mineralogische Spezies," *Gesnerus*, 1984, 41:257–268.
22 François Sulpice Beudant, "Recherches tendantes à déterminer l'importance des formes cristallines et de la composition chimique dans la détermination des espèces minérales," *Annales des mines*, 1817, 2:1–29.
23 François Sulpice Beudant, *Traité élémentaire de minéralogie* (Paris, 1824), p. 246; 2nd ed., 2 vols. (Paris, 1830–1832), Vol. I, p. 401.

polymorphism for granted. In 1827 Mitscherlich's French colleagues ended the debate on isomorphism with the symbolic act of electing its discoverer to the Academy of Sciences as a corresponding member.[24] Berzelius's influence on this relatively quick and smooth recognition of Mitscherlich and his work cannot be perceived in quantitative terms, but it must have been considerable.

24 On 31 December 1827 Mitscherlich became "Correspondant" for the section "Minéralogie"; on 14 June 1852 he became "Associé étranger"; Lacroix, "Tentative," p. 206.

8

Berzelius, the Dualistic Hypothesis, and the Rise of Organic Chemistry

JOHN HEDLEY BROOKE

Introduction

The career of Berzelius coincided with a period in which a new science emerged. By the time of his death, organic chemistry had burst the bounds of an earlier conception in which it had been presented as the science of animal and vegetable products. As the chemistry of carbon compounds, it had expanded its domain to include a bewildering variety of artificial derivatives that might, or might not, feature in natural physiological processes. For 30 years and more, Berzelius presided over this process of differentiation, suffering the fate of many great scholars in both the arts and sciences. From having been the pioneer who in 1814 first applied the new atomic theory to organic compounds, he took his place, through the *Jahresberichte,* as commentator and adjudicator of the work of a younger generation of experts (notably those emerging from Justus Liebig's school at Giessen), only to find his methods, the style of his work, and ultimately his authority overtaken by innovations he was powerless to suppress. A balanced assessment of his role in defining and patrolling the boundaries of the new science must be true to each of these phases in his career. Unfortunately, however, Berzelius's reputation as an organic chemist has suffered through an undue concentration on that last, infelicitous phase when he fought his rearguard action against the French – against the substitution theories of Jean Baptiste André Dumas (1800–1884), Auguste Laurent (1807–1853), and Charles Frédéric Gerhardt (1816–1856) who were threatening to undermine the edifice of organic theory that Berzelius had constructed on the foundations of electrochemical dualism. Not even J. R. Partington could refrain from accusing Berzelius of reck-

Significant parts of the present study first appeared in "Chlorine substitution and the future of organic chemistry: methodological issues in the Laurent/Berzelius correspondence (1839–1843)," *Studies in History and Philosophy of Science,* 4 (1973), 47–93, and is reprinted with permission from Pergamon Press PLC.

lessness, dogmatism, and obstinacy in his forlorn attempt to hold the fort.[1]

One of the objectives of this chapter is to offer a more sympathetic reconstruction of Berzelius's activity in the organic sphere. In accordance with an earlier analysis,[2] particular attention is paid to the resilience with which he fought for a principle of method that had been invoked, particularly during the 1830s, to justify the extension of dualistic concepts from mineral chemistry, where since Lavoisier's treatment of acids, bases, and salts they had been endemic, to the formulation of organic compounds, where they were eventually (though by no means initially) found wanting. The conservative stance that Berzelius adopted toward the claim that chlorine could take the place and play the role of hydrogen in organic compounds was not the reaction of one who only saw his electrochemical theory under threat. It was also the reaction of one with a profound commitment to the unity of chemical theory who rightly believed that a dualistic approach to the constitution of organic compounds had been successful – too successful to be abandoned.

This methodological dimension to Berzelius's chemistry has now been more widely recognized,[3] and it was certainly recognized by Berzelius himself. In a letter to August Laurent, one of his most vigorous assailants, Berzelius identified the nub of their disagreement. In the knowledge that Laurent's theoretical concepts, and particularly that of a stable hydrocarbon nucleus in which substitution could occur with retention of properties, had their origin in his study of naphthalene and its derivatives, Berzelius observed:

> You are endeavouring to reform the theory of inorganic chemistry according to ideas you have derived from your experience with organic compounds. In my chemical studies I have preferred the diametrically opposite route, that of basing speculations about organic composition on the theoretical ideas more or less established for inorganic compounds.[4]

Laurent's inversion of a principle of method that Berzelius had presented as a sine qua non of progress in the science was simply too much to bear.

---

1 James R. Partington, *A History of Chemistry*, 4 vols. (London: Macmillan, 1961–1970), Vol. IV, pp. 369 and 371.

2 John H. Brooke, "Chlorine Substitution and the Future of Organic Chemistry: Methodological Issues in the Laurent-Berzelius Correspondence," *Studies in the History and Philosophy of Science*, 1973, 4:47–94.

3 Trevor H. Levere, *Affinity and Matter: Elements of Chemical Philosophy, 1800–1865* (Oxford: Clarendon Press, 1971), pp. 155–157; Colin A. Russell, *The Structure of Chemistry* (Milton Keynes: Open University Press, 1976), unit 1, pp. 20 and 42.

4 Berzelius to Auguste Laurent, June 1844, in *Bref*, Vol. III:ii, p. 208.

It was as if organic chemistry were asserting a complete and voracious autonomy to which, in Berzelius's opinion, it was not yet entitled.

To evaluate Berzelius's contribution to organic chemistry, it is therefore necessary to assess the consequences of his methodology, with its insistence on analogical argument from the inorganic to the organic domain. A difficulty, however, arises at the outset, because it might be asked whether such principles of method actually *have* any consequences, any real effects on the practice of science. Does the fact that Berzelius was justifying himself in his letter to Laurent not suggest that his discourse on method was fulfilling a purely rhetorical function, offering a retrospective rationalization of certain forms of chemical practice, rather than delineating a method of inquiry that had been consciously applied in regulating that practice? The question is important because an increasingly sophisticated literature on the social history of methodology has emphasized the diversity of rhetorical functions that method-discourse can fulfill. Much more is now known about the manner in which methodological prescriptions have assisted in the popularization of controversial theories, in the promotion of particular types of science as normative, in accounting for the erroneous character of rival science, in demarcation disputes between different disciplines, and in cultivating images of scientific inquiry likely to impress the public.[5] Where such rhetorical functions are emphasized, one may find method-discourse relegated to a category of "mythic speech" in which reported applications of a formalized methodology become phantom redescriptions, the very purpose of which generates linguistic structures that cannot mesh with, and certainly cannot determine, scientific practice.[6] The objection might, therefore, be that Berzelius's discourse on method is too slender a reed on which to build a case for revision.

There is, however, a crucial distinction between rhetoric and mere rhetoric, and the objection would be stronger if Berzelius had promoted his analogy principle only in self-defense and under duress. The fact is that he promoted it earlier in his career and in contexts where he was prepared to approve the work of others, even potential rivals. In his *Jahresbericht* for 1828, as he reviewed the hypothesis of Dumas and Pierre François Guillaume Boullay (1777–1858), which made the hydrocarbon radical C^2H^4 the base of both ether and alcohol, he was not slow to criticize, but with an eye on a possible analogy between C^2H^4 and ammonia he nevertheless pronounced the theory an important attempt to "understand the mutual proportions of the elements by attaching them

5 John A. Schuster and Richard R. Yeo, eds., *The Politics and Rhetoric of Scientific Method* (Dordrecht: Reidel, 1986).

6 John A. Schuster, "Cartesian Method as Mythic Speech: A Diachronic and Structural Analysis," in Schuster and Yeo, eds., *Politics*, pp. 33–95, especially pp. 59–79.

to the formulas of other well known compositions."[7] When he wrote to Justus Liebig (1803–1873) in 1833, claiming that the only safe starting point for theories of organic composition was the projection of analogies from inorganic chemistry,[8] he was no doubt affirming his control over the science, but he was at the same time defining a procedure that could mesh with practice. Indeed, one of the reasons why Berzelius's methodology deserves attention is that it would seem to constitute an excellent case study in the overlap (rather than mutual exclusivity) of substantive and rhetorical functions. The analogical projections were so visible in the theoretical content of his science that the methodology that sustained them has to be taken seriously.

That method-discourse need not be entirely epiphenomenal with respect to scientific practice is one of the subsidiary themes of this chapter. The analysis will begin by suggesting that the regulative character of Berzelius's methodology emerged from a dialectical process in which a strong predisposition toward the unity of chemistry encountered empirical data that, while sufficiently supportive to maintain the drive toward unification, were yet sufficiently noncompliant to set up a creative tension in Berzelius's mind – a tension that found expression in the prescription that organic compounds should be modeled as far as possible on those of mineral chemistry, but without prejudice to the outcome.

Quantitative Organic Chemistry and the Origins of a Methodology

Since Berzelius's earliest work in the sphere of animal chemistry is the subject of a separate chapter by Alan Rocke,[9] the starting point here will be the paper of 1814 in which Berzelius reported his success in applying atomic concepts to organic combination.[10] Two inferences drawn by Rocke are, however, crucial to an understanding of Berzelius's intellectual development. First, in the early publications on animal chemistry, there is a clear preference for mechanist and materialist explanations of physiological phenomena. While careful to distance himself from extreme forms of materialism that might lead to the denial of an immortal soul, Berzelius was quite prepared to say:

7 Berzelius, *JB*, 1829, 8:287–297; Claus Priesner, "Spiritus Aethereus – Formation of Ether and Theories on Etherification from Valerius Cordus to Alexander Williamson," *Ambix*, 1986, *33*:129–152, on p. 135.
8 Berzelius, "Ueber die Constitution organischer Zusammensetzungen," *Annalen der Chemie*, 1833, 6:173–176.
9 Alan J. Rocke in this volume, Chapter 5.
10 Berzelius, "Experiments to Determine the Definite Proportions in Which the Elements of Organic Nature Are Combined," *Annals of Philosophy*, 1814, 4:323–331, 401–409; 1815, 5:93–101, 174–184, 260–275; cf. Evan M. Melhado, *Jacob Berzelius: The Emergence of His Chemical System* (Madison and Stockholm, 1979), Ch. 6.

All of the brain's functions are caused by and depend on all of the same physical and chemical laws as the other functions of the body. . . . As absurd and as paradoxical as this may seem . . . nonetheless, our judgement, our memory, our thoughts, indeed, all of the brain's functions, are fully as much organic-chemical processes as are those of the stomach, intestines, lungs, glands, etc.[11]

Thus, even before he was in a position to test the applicability of atomic theory to organic compounds, Berzelius had been predisposed toward the view that the same laws would obtain in both organic and inorganic chemistry. There was no need, in principle, to hypothesize other forces peculiar to living systems.[12] Second – and Rocke here refers to an element of irony – at just the time he set out to assimilate organic and inorganic products, he discovered some differences between them. The creative tension generated by this collision is clearly visible in the 1814 paper, for Berzelius was at pains to show that there were fundamental resemblances between organic and inorganic compounds, even if a complete assimilation could not yet be achieved.

At least four such resemblances were underlined. First, the results of quantitative analysis for both organic and inorganic compounds could be interpreted in accordance with atomic theory. The applicability of the atomic theory to organic compounds had by no means been self-evident, for in an earlier paper Berzelius had been flummoxed by an empirical formula of $C^{27}H^1O^{18}$ for oxalic acid that had cast doubt on the utility of atomic models in exploring organic composition.[13] It had been difficult to see how one atom of hydrogen could be linked to 45 other atoms. Berzelius's apprehension had been sufficiently marked to catch the eye of John Dalton, who came up with a possible solution: "Were it a matter of necessity, an atomist might conceive 1 atom of hydrogen surrounded by 9 of carbon, and the compound globule to have 18 atoms of carbon oxide adhering to it."[14] The problem with that solution was that it blithely assumed that organic compounds could be resolved into preformed groups of elements that, as far as Berzelius was concerned, was one of the central points at issue. As it transpired, it was not a matter of necessity, and a subsequent analysis produced a less formidable formula, $C^{12}H^1O^{18}$. With the removal of such difficulties, Berzelius was now in a position to argue that "in the present state of our knowledge the corpuscular theory is the

11 Cited by Rocke in this volume, Chapter 5, p. 118, text accompanied by note 46.
12 In his letter to C. A. Agardh of 22 November 1831 (*Bref*, Vol. IV:iii, pp. 71–72), also cited by Rocke and Sven-Eric Liedman (Chapter 2) in this volume, Berzelius held that the difference lies not in the forces but in the circumstances in which they act.
13 Berzelius, "Cause of Proportions," pp. 447–450.
14 John Dalton, "Remarks on the Essay of Dr. Berzelius on the Cause of Chemical Proportions," *Annals of Philosophy*, 1814, 3:174–180, on p. 178.

only one which puts it in our power to explain the composition of organic atoms in a satisfactory manner."[15] No longer disturbed by the larger numbers appearing in organic formulas, he was resigned to the fact that the mechanical structures of organic compounds might differ from those of the inorganic realm.[16] Of the applicability of the atomic theory to both domains, however, there was no longer any doubt.

Second, the very title of the 1814 paper alluded to an underlying resemblance: "Experiments to Determine the Definite Proportions in Which the Elements of Organic Nature Are Combined." The applicability of the atomic theory presupposed that in a specific organic compound, as in mineral compounds, the parts by weight of each element were fixed and definite. Because more complex organic materials had yielded variable results on analysis, there was a larger issue at stake – namely, whether the concept of definite proportions could be used to distinguish between organic compounds and the more amorphous composites that were a feature of living systems. It was in just such a situation that the regulative character of an analogy principle could assert itself. As Berzelius himself wrote: "It is evident that the existence of determinate proportions in inorganic bodies leads to the conclusion that they exist also in organic bodies."[17]

Third, the same self-conscious extension of concepts from one realm to the other is visible in the use Berzelius made of a law of combination that he considered more basic than the law of multiple proportions developed by Dalton. Berzelius had satisfied himself that when two inorganic oxides combined with each other, the oxygen in the one was always a whole number multiple of that in the other.[18] It was by means of that law, Berzelius declared, that "we can hope to penetrate into the secrets of organic composition."[19] The same predisposition was at work. Once that law could be demonstrated for organic compounds, the theory of chemical proportions could be "considered as established." Not all the secrets were penetrated in this way, but Berzelius continued to use his oxide law, both to fix the empirical formulas of organic acids (by saturating them with bases of known oxygen content)[20] and to reject spurious analogies between the alkaloids and inorganic oxides.[21]

Fourth, his belief – soon to be modified – that all mineral acids were oxides, their hydrates being of the form $[R \cdot O + H^2O]$, encouraged tentative steps in the direction of representing organic acids as oxides of

15 Berzelius, "Experiments," 1815, pp. 274–275.   16 Ibid., p. 98.
17 Berzelius, "Experiments," 1814, p. 323.
18 On the genesis of this oxide rule see Melhado, *Berzelius*, pp. 179–184.
19 Berzelius, "Experiments," 1815, p. 128.
20 Berzelius, *Traité de Chimie*, Vol. V (Paris, 1831), p. 16.
21 Ibid., Vol. VI, 2nd ed. (Paris, 1850), p. 11.

compound hydrocarbon radicals. Such steps were, for Berzelius, rarely more than a convention to draw out the stoichiometric analogies between inorganic and organic acids. They did, however, become his official line on the organic acids, and they led him to the view that the only significant difference between organic and inorganic acids lay in the complexity of the organic radicals. Never one to throw caution to the winds, he was still insisting five years later that the hypothesis of hydrocarbon radicals was highly conjectural, not least because the independent existence of the radical outside its compound was problematic.[22] Nevertheless the analogies in composition between organic and inorganic acids helped to build a bridge between the two domains, even if in the long run a more sophisticated concept of acidity was required than that which Lavoisier had bequeathed.[23]

In his 1814 paper, Berzelius had made some headway in his expected assimilation of organic and inorganic compounds; but there were problems and it was not in his character to shirk them. In four other respects, a complete analogy between the constitution of organic and inorganic compounds was frustrated. To appreciate the first of these it is necessary to know that, differing from Dalton, Berzelius would not at first entertain the possibility of formulas such as A^2B^3, A^2B^5, A^3B^4 for inorganic compounds. In Berzelius's own words, "it was contrary to sound logic" to represent a compound atom of the first order by a formula in which neither element appeared as a single atom.[24] One's sympathies may lie with Dalton, who, unimpressed by sound logic, could see no objection to A^2B^2, A^2B^3, . . . And yet Berzelius would reply: Why represent the oxides of iron as FeO, Fe^3O^4, Fe^2O^3 when they could more simply be written FeO^6, FeO^8, FeO^9? There is no need to pursue this essentially aesthetic issue,[25] except to note that in Berzelius's eyes, "The two circumstances, that, first, the compounds of the first order are always binary, and that, secondly in all these combinations at least one of the constituents enters only in the quantity of a *single* atom or volume, constitute the exclusive characters of inorganic nature."[26] The problem was that in contemplat-

22 Colin A. Russell, "The Electrochemical Theory of Berzelius," *Annals of Science*, 1963, 19:117–145, on pp. 117 and 127.

23 For an introduction to theories of acidity in the early nineteenth century, with particular reference to Davy's critique of Lavoisier, see John H. Brooke, "Davy's Chemical Outlook – The Acid Test," in Sophie Forgan, ed., *Science and the Sons of Genius: Studies on Humphry Davy* (London: Science Reviews, 1980), pp. 121–175.

24 Berzelius, "Cause of Proportions," pp. 447–450.

25 On this dispute between Berzelius and Dalton, see Colin A. Russell, "Berzelius and the Development of the Atomic Theory," in *John Dalton and the Progress of Science*, Donald S. L. Cardwell, ed. (Manchester: Manchester University Press, 1968), pp. 259–273; and Melhado, *Berzelius*, pp. 270–278.

26 Berzelius, "Experiments," 1814, p. 325.

ing organic nature reduction to unity for at least one of the components was prevented by empirical formulas of considerable complexity.

Indeed, the pressure was toward increasing the variables for each element, the better to account for the most striking phenomenon of all – that such an immense diversity of organic products should be spun from the parsimonious resources of four elements: carbon, hydrogen, oxygen, and nitrogen. The greater mobility and instability of so many organic compounds also invited explanation in terms of the larger number of atoms of each constituent than was the norm in the mineral realm. As the British chemist Andrew Ure was to observe in 1822:

> All the elementary principles of organic nature may be considered as deriving the peculiar delicacy of their chemical equilibrium, and the consequent facility with which it may be subverted and new modelled, to the multitude of atoms grouped together in a compound. On this view, none of them should be expected to consist of a single atom of each component.[27]

There were, then, considerations that pulled in opposite directions, and Berzelius experienced the tension in 1814. It should be noted, however, that this particular tension could be resolved by releasing the strictures he had placed on *inorganic* combination. And that is precisely what he later did. The increasing numbers of compounds, such as the oxides of phosphorus, which could be conveniently represented in the form A_2B_3, A_2B_5, together with the self-consistent application of his rule that equal volumes of gases under standard conditions contain the same number of atoms, forced him to change his mind. A decade later and he was perfectly happy to write the oxides of nitrogen as $2N + O; N + O; 2N + 3O; 2N + 5O; 3N + 6O$, in defiance of his earlier rule.[28]

There was a second respect in which the patterns of combination characteristic of simple binary compounds pointed toward a dissimilarity between organic and inorganic species. This was at the level at which one could ask whether recurrent, discernible patterns of combination existed at all among organic compounds. However much Berzelius was prepared, or not prepared, to release his strictures on the patterns for inorganic compounds, he had no doubt that certain patterns in the range AB, AB_2, . . . to AB_{12} commonly existed,[29] while others such as A_2B_4, A_2B_5, A_2B_6 were inadmissible either in principle or in practice. In other words,

---

27 Andrew Ure, "On the Ultimate Analysis of Vegetable and Animal Substances," *Philosophical Transactions of the Royal Society*, 1822, 112:457–482, on pp. 468–469.

28 Partington, *History*, Vol. IV. p. 212; cf. Melhado, Chapter 6, in this volume, text accompanied by note 58.

29 If spherical atoms of identical size are assumed, 12 is the maximum number that could be packed around one.

in the inorganic sphere there were limits to the combining ratios in which the constituent elements appeared.

Part of what it meant for Berzelius to construct a theory of chemical proportions was that some explanation be found for the source of the constraints. Putting the matter anachronistically, it was as if Berzelius was on the scent of some concept of valency, without having the wherewithal to articulate or consolidate it. But the question was whether such patterns of restriction were discernible among the empirical formulas of organic substances. And in 1814 the answer appeared to be "no." Formulas of the general type $C^xH^yO^z$ were found that not only evinced very large values for x, y, and z but that also seemed to defy any regulation as to what might be legitimate values of those parameters.

The situation was still essentially the same in 1819, when Berzelius published his *Essay* on chemical proportions. Hence his statement that "The laws that limit the combinations of elementary atoms in organic nature differ considerably from those that regulate inorganic nature and permit such a multiplicity of combinations that it may be said that no determinate proportions exist."[30] Of all Berzelius's pronouncements this continues to be the most commonly misunderstood. Scholars have not only caught the whiff of what they take to be a chemical vitalism, but they have even imagined that Berzelius was here denying what, in 1814, he had taken such pains to affirm – namely, that there were definite proportions in one and the same organic compound.[31] All he was in fact denying was that there were characteristic patterns in organic formulas governing the ratios in which their few elements combined. There was no denying, however, that this was a difference between the two realms. Berzelius scrupulously noted it, but did his best not to emphasize a gulf but to establish a continuum between them. In 1814 this effort took the form of an assertion to the effect that in organic nature the composition and the laws of proportion became increasingly complex as the number of elements increased.[32] Later, he would single out calcium phosphate as an exemplar of that continuum. Though it was an inorganic salt, it exhibited a combining ratio unique in inorganic chemistry: eight atoms of calcium oxide united to three of phosphoric acid. But that unique ratio was perfectly intelligible, Berzelius suggested, in the light of the *organic* function of the compound.[33]

30 Berzelius, *Essai sur la Théorie des Proportions Chimiques* (Paris, 1819), p. 40 [my translation].
31 Cf. Jean Jacques, "Le vitalisme et la chimie organique pendant la première moitié de XIXe siècle," *Revue d'histoire des sciences*, 1950, 3:32–66, on p. 49. The applicability of Proust's law of fixed composition for chemical compounds was not the issue in 1819.
32 Berzelius, "Experiments," 1814, p. 325.
33 Berzelius, *Théorie des proportions chimiques* (Paris, 1835), p. 23.

In his analyses of the vegetable acids, Berzelius had found no example of a compound without oxygen as a constituent. Looking for a useful generalization, he inferred that "all organic bodies contain oxygen united to more than one combustible radicle. . . ."[34] The significant point is not so much the claim for the omnipresence of oxygen, but rather the holistic implications of the view that in organic compounds the oxygen was, as it were, shared by two other components. The implication was that the compound could not be separated into proximate constituents without destruction of the whole. Superficially at least this implication contrasted with the state of affairs in mineral chemistry where the acid/base dualism of the electrochemical theory made it not inappropriate to envisage a salt as containing the neutralized acid and base within it.

In a nutshell, the problem was this: Whereas inorganic acids and bases were believed to contain only two elements (Na^2O, SO^3, . . .), all organic products appeared to contain at least some carbon, hydrogen, and oxygen. Had it been possible in 1814 to produce an organic compound in the laboratory from two binary parts, just as a salt was produced from acid and base, an analogy might have been sustained; but Berzelius felt obliged to report that "chemical experiments on organic substances have shown us that these combinations of oxygen with two or more radicles cannot be considered as composed of two or more binary bodies."[35] Coloring matter of the blood, for example, ought to be considered a compound of carbon, hydrogen, nitrogen, phosphorus, calcium, and iron, *all* combined with a portion of oxygen common to them all.[36] With such a homogeneous image of an organic substance, the prospects for introducing an electrochemical dualism into the organic kingdom looked none too bright.

The fourth of the dissimilarities to which Berzelius drew attention in 1814 he actually described as an "essential difference" between organic and inorganic nature. This consisted in the "electrochemical modification of the organic products, which does not appear to depend immediately on . . . the original modifications of the elementary substances."[37] Here, Berzelius was repeating an earlier statement to the effect that "the chief condition for organic formation seems to be an electrochemical modification in the elements which differs from that which they normally have in inorganic nature."[38]

It is statements such as these, scattered throughout his works and still reiterated in the last edition of his textbook, that have given rise to the view that Berzelius was, at heart, a vitalist. In earlier papers I have argued

34 Berzelius, "Experiments," 1814, p. 325.   35 Ibid.
36 Ibid., p. 328.   37 Ibid., pp. 328–329.
38 Berzelius, "Essai sur la nomenclature chimique," *Journal de physique,* 1811, 73:253–286, on pp. 260–261, cited by Russell, "Electrochemical Theory," p. 136.

against this interpretation, which rests on a failure to clarify what Berzelius meant by "vital force."[39] He did not doubt that there were peculiar powers at work in a living organism, nor did he doubt that they emerged from the complexity of material organization. In his contribution to this volume, Alan Rocke has reached a similar conclusion, suggesting that insofar as Berzelius had a coherent position it was that of the German exponents of what Timothy Lenoir has called "vital materialism."[40] It was, of course, open to Berzelius's contemporaries to seize his holistic view of organic products if they wished. Johannes Müller gave it an overtly vitalistic interpretation in his *Elements of Physiology*.[41] But to suggest that Berzelius in 1814 was looking for evidence that would support vitalism over a materialist philosophy would be to miss both the subtlety of his position and that drive toward a unification of chemical theory that he had already made explicit. One can see why historians have taken Berzelius's remarks about electrochemical modification to imply that he favored the operation of an additional, independent vital force; but a closer analysis of his published work reveals their mistake.

In the first place, the context of Berzelius's references to electrochemical modification was not one in which the question of a vital force, or even that of organic synthesis, was uppermost. The point at issue was whether the properties of an organic compound could be predicted from the number and properties of its constituent elements. Gay-Lussac and Thenard had formulated certain rules that ostensibly allowed one to predict the properties of an organic compound merely from the proportions of its elements. For non-nitrogenous compounds, an oxygen-to-hydrogen ratio that exceeded that of water guaranteed their acidity, irrespective of their carbon content. If the O:H ratio were identical to that of water, the compound would be neutral, analogous to sugars and gums.[42] The attraction of this numerical determinism was considerable, and it may help

39 John H. Brooke, "Wöhler's Urea and Its Vital Force? A Verdict from the Chemists," *Ambix*, 1968, 15:84–114, especially p. 85; John A. Brooke, "Organic Synthesis and the Unification of Chemistry: A Reappraisal," *British Journal for the History of Science*, 1971, 5:363–392, especially pp. 380–381.

40 Timothy Lenoir, *The Strategy of Life: Teleology and Mechanics in Nineteenth Century German Biology* (Dordrecht: Reidel, 1982).

41 Johannes Müller, *Elements of Physiology*, tr. William Baly, Vol. I (London, 1838), pp. 1–38. See also Edward Benton, "Vitalism in Nineteenth-Century Scientific Thought: A Typology and Reassessment," *Studies in History and Philosophy of Science*, 1974, 5:17–48, especially pp. 29–35.

42 The rules were summarized in Louis J. Thenard, *Traité de chimie*, 4th ed. (Paris, 1824), p. 560. On the collaboration between Thenard and Gay-Lussac, see Maurice P. Crosland, *Gay-Lussac: Scientist and Bourgeois* (Cambridge: Cambridge University Press, 1978).

to explain some of the initial resistance to the concept of isomerism. Berzelius, though, was not so readily taken in. In a pointed refutation he observed that benzoic acid, which contained only the fifth part of its weight of oxygen, was an acid, whereas gum arabic and sugar, which contained nearly half their weight of oxygen, were not acids. His conclusion was uncompromising: "I have not hitherto been able to perceive that either the number of volumes, or their relation to each other, determined any thing respecting the electro-chemical properties of ternary (organic) oxides."[43] Consequently, when he confessed that "it is impossible to determine in organic bodies from the elements and proportions of which they are composed whether they be acid or not,"[44] this statement should not be construed as an admission of vitalist belief, but simply as a critique of the numbers game played by Gay-Lussac and Thenard.

Moreover, the conclusion that some modification of the electrical properties of the elements was effected "in vivo" was forced on Berzelius by the evidence. It was not deduced from a vitalist premise. From his experience with inorganic oxides, he surmised that the higher the oxide, the more acidic it would be. Transferring this inference to organic chemistry, one would expect that of the three acids CO (carbon monoxide), C^2O^3 (oxalic acid), and CO^2 (carbonic acid), the last should be strongest. Berzelius was never to solve the mystery as to why it was that oxalic acid was so much stronger than the other two, but this fact immediately implied some modification of the carbon.

From such empirical data, Berzelius was led to the view that "at the instant of the formation of each ternary and quaternary oxide in organic nature, its elements receive in combining a new electrochemical modification, on which their chemical properties chiefly depend."[45] It is difficult to know exactly what Berzelius had in mind, but, in contradistinction to the view of at least one commentator,[46] he was not denying that the properties did still depend on the (albeit new) electrochemical character and the proportions of the elements. There was certainly no question of an extrachemical factor lurking within an organic compound. Current failure to achieve an artificial synthesis could obviously have been used to support the idea of an instantaneous modification produced exclusively in vivo, but that does not appear to have been what Berzelius had in mind. In fact, he was later quite explicit in denouncing such an association of ideas. Writing to C. A. Agardh in 1831, he insisted that "any proof based on inability is no proof at all." To suppose that the elements

43 Berzelius, "Experiments," 1814, pp. 328–329.   44 Ibid.
45 Ibid., p. 329.
46 A. Costa, *Michel Eugène Chevreul* (Madison: University of Wisconsin Press, 1962), p. 25.

were imbued with other fundamental forces in organic nature than obtained in the inorganic realm was "an absurdity."[47]

Such differences as there were between organic and inorganic compounds, Berzelius informed Agardh, were due to the different conditions under which synthetic forces worked. But the fact that the conditions prevailing in organic nature were not yet understood gave no reason to adopt other forces. On one reading of such remarks, the references to electrochemical modification did not even imply the improbability of organic synthesis in the laboratory. Was there any reason in principle why the chemist might not eventually re-create those special conditions in vitro? As Berzelius's contemporary, Eilhard Mitscherlich, was soon to write: The problem exercizing the chemists of his day was to find the appropriate conditions.[48] One point is certain. In the 1814 paper, Berzelius ascribed the electrochemical modification to the working of the nervous system, not to an autonomous vital force.[49] As for the working of the nervous system, he frequently wrote as if he expected that an electrical mechanism would be forthcoming. A few years later, in 1822, he reflected that "the more the phenomena of organic nature convince us that electricity is the final cause of chemical action, the more it becomes apparent that it also determines the processes in the animal kingdom which at least resemble those of inorganic nature, and the more justified we are in supposing that the working of the nervous system depends chiefly on this force."[50]

The differences between organic and inorganic compounds that Berzelius reported in 1814 were not, then, intended to give *Lebensraum* to a *Lebenskraft*. Looking ahead to the *Essay* of 1819, one may say that the vital force was to be conspicuous by its absence.[51] And looking still further ahead, one may observe that Berzelius would draw chemists' attention to the fact that many compounds exhibiting isomerism contained elements that themselves exhibited different allotropic forms. It was not impossible, he speculated, that the different modifications of the organic

---

47 The significance of this letter for an appreciation of Berzelius's mature position was first appreciated by Bent S. Jørgensen, "More on Berzelius and the Vital Force," *Journal of Chemical Education,* 1965, 42:394–396.

48 Eilhard Mitscherlich, *Élémens de chimie,* tr. M. B. Valérius, Vol. II (Brussels, 1835–1836), p. 3. Once Berzelius acknowledged, as he later did, that artificial production of organic compounds had been achieved, he drew the conclusion that it *would* be going too far to suggest that a vital force was capable of completely effacing the fundamental powers of matter, a statement that could, however, be read in more than one way; Berzelius, *Traité,* 2nd ed. (Paris, 1848–50), Vol. V, p. 5.

49 Berzelius, "Experiments," 1814, p. 329.   50 Berzelius, *JB,* 1822, 1:116.

51 Colin A. Russell, "Introduction to the Reprint Edition [of Berzelius's *Essai sur la théorie des proportions chimiques*]" (New York and London: Johnson, 1972), pp. [v]–[xlix], on p. [xxxii].

elements in vivo corresponded to their elementary allotropic modifications[52] – in which case any appeal to vital forces would be superfluous as well as misguided.

The four differences could not, however, be ignored, and they prevented Berzelius from drawing any dogmatic conclusion about the relationship between organic and inorganic substances. The exploratory character of his methodological principle (at once both regulative and tentative) was therefore a realistic response to the conjunction of at least three different pressures: a predisposition to show that the laws governing the chemistry of organic and inorganic compounds were the same, the realization that there might yet be something distinctive about organic products, and the burning necessity to produce some conjecture as to how the elements might be arranged in organic compounds. In making such conjectures some of Berzelius's contemporaries were more adventurous than he was prepared to be. Indeed, one of the grounds on which a more positive assessment of *Naturphilosophie,* in its bearing on chemistry, has been urged is the perspicacity with which German chemists such as Karl Kastner and Johann Wolfgang Döbereiner were postulating reaction mechanisms and schemes of classification before the dualistic categories of Berzelius's electrochemistry were fully operative in rationalizing organic formulas.[53] During the ensuing 25 years, from 1814 to 1839, it was nevertheless the methodology of Berzelius that came to fruition. His cautious insistence that organic products be modeled on those of the inorganic domain was to have a more enduring effect on the development of organic chemistry than the sparkling, but usually ephemeral, speculations of the *Naturphilosophe.*

The Extension of Dualism into Organic Chemistry

So readily did the acid/base dualism of Lavoisier's chemistry lend itself to interpretation in electrical terms that, in the hands of Berzelius, it acquired an aura of such sophistication that for many observers it almost appeared as if inorganic chemistry had been finished. Every new compound, and its properties, could be accommodated by a set of ideas that Gay-Lussac and Dumas described in 1833 as *"parfaitement arrêté et dé-*

---

52 Berzelius, "Om allotropi hos enkla kroppar, såsom en af orsakerna till isomeri hos deras föreningar" [On the Allotropy of Simple Bodies . . .], KVA, *Handlingar, 1843:*1– 18; read to the Academy of Sciences at Stockholm, 13 September 1843. A French translation appeared in *Revue scientifique et industrielle de Quesneville,* 1843, *15*:137, and an English version in *Taylor's Scientific Memoirs,* Vol. IV, pp. 240–252.

53 Reinhard Löw, "The Progress of Organic Chemistry during the Period of German *Naturphilosophie* (1795–1825)" *Ambix,* 1980, 27:1–10.

fini."[54] Berzelius's concept of bipolar atoms, his distinction between the intensity of polarization (how much charge might be concentrated at each pole) and the overall charge on the atom (the difference between the charges at each pole), provided a degree of flexibility in accounting for the observed affinities between specific elements and between their compounds.[55] The theoretical superstructure was aesthetically very compelling. Every salt could be divided into acid and base. Every acid and every base could, in turn, be written as a compound of two elements, or as the hydrate of an oxide – in which case the binary pattern was retained. The question that would not go away was whether the same binary patterns could not be imposed on organic combinations. During the 1820s and 1830s it became increasingly plausible to suggest that they could.

Despite the reservations Berzelius had voiced in 1814, one is struck in retrospect by the ease with which organic compounds could be viewed through the spectacles of dualism. Certainly the major classes of organic substances could be pressed into the mold without obvious violation. Organic acids, as Berzelius had already implied, could be formulated as material analogues of inorganic relatives such as sulfuric. The same formula $[XO^n + H^2O]$ could be applied to both, the only difference being that for an organic acid the radical X would not be a single element, like sulfur, but a compound of carbon and hydrogen. Organic bases, too, were just as susceptible to dualistic interpretation. A series of them, which he called cinchonine, kinine, and aricine, were written by the French chemist Pelletier in the form $R + O, R + 2O, R + 3O$, where R was a composite group analogous to an inorganic element.[56] The existence of hybrid salts such as the phosphovinates and sulfovinates helped to corroborate the dualistic construction of organic acids and bases. Other classes of organic compound lent themselves equally well to the divisions of dualism. The fact that sulfuric acid appeared to dehydrate ethyl alcohol in the production of ether encouraged the idea that alcohol might be a simple hydrate of ether. It was then perfectly possible to represent esterification reactions in conformity with a dualist rubric:[57]

$$[C^4H^{10}O + H^2O] + [C^4H^6O^3 + H^2O] =$$
$$[C^4H^{10}O + C^4H^6O^3] + 2H^2O$$

54 Joseph L. Gay-Lussac and Jean B. A. Dumas, "Rapport fait à l'Académie des sciences, sur un mémoire de M. Pelletier intitulé Recherches sur la composition élémentaire de plusieurs principes immédiats des végétanx," *Journal de Pharmacie*, 1833, 19:93–99, on pp. 93–94.

55 Russell, "Introduction," pp. [xl]–[xli]; Melhado, *Berzelius*, p. 152–164.

56 Gay-Lussac and Dumas, "Rapport," p. 97.

57 Justus von Liebig, *Traité de chimie organique*, tr. Charles Gerhardt, Vol. I (Paris, 1840), p. 314.

That esters could be so readily hydrolyzed to regenerate the acid and the alcohol seemed to give a remarkable confirmation of the analogy with inorganic salts. Acids, bases, and esters all complied with the dualism borrowed from mineral chemistry. So did the halides, which during the 1830s were classed along with the esters under the generic "ether." What we know as ethyl chloride was then known as hydrochloric ether, the production and hydrolysis of which could be envisaged along parallel lines to the production and hydrolysis of common salt. Schoolboy chemists still imagine the alkyl halides to be ionic; and there are some, like the tertiary halides, that really are. By 1830, Dumas had convinced himself that amides could be treated as analogs of the halides, and so they too fell into line.[58] Dumas's formula for oxamide was $[C^4O^2 + Az^2H^4]$, and any doubt as to the propriety of using the hypothetical radical Az^2H^4 was dispelled by the existence of sodamide and potassamide in which the same radical was clearly united with a metal. Finally the fats: They, too, conformed to the dualist pattern; for, as Michel Eugène Chevreul, who pioneered their study, reported, "The fats, in many respects, are quite analogous to the esters, which are considered to be compounds of acids and alcohol."[59]

The extension of dualism into organic chemistry was not, however, so straightforward or so unequivocal in its results as to stifle controversy. Writing in 1837, Liebig and Dumas confirmed that one of the grandest questions in natural philosophy had been *how* to apply the laws of inorganic chemistry to the organic field.[60] The problem during the 1820s and 1830s was that in the details of that *how* more than one projection was always possible. Granted that the basic pattern for inorganic chemistry was the combination of elements two by two, and their compounds two by two, there were at least two ways in which this pattern could be drawn into organic combination. In the first place, organic compounds could likewise be represented as binary compounds of their *elements,* carbon, hydrogen, oxygen, and nitrogen. Thus Gay-Lussac and Dumas wrote alcohol as a hydrate of a hydrocarbon; Mitscherlich formulated benzoic acid as $(C^{12}H^{12}) + 2\ (CO^2)$; William Prout saw the sugars as hydrates of carbon; and Dumas represented oxamide by $(C^4O^2) + (Az^2H^4)$. But, in the second place, organic compounds could always be envisaged as binary compounds of composite *groups* of elements – groups, or radicals, which as analogs of the inorganic elements might contain as many

---

58 Dumas, "Sur l'oxamide, matière qui se rapproche de quelque substances animales," *Annaels de chimie,* 1830, 44:129–143.

59 Cited by P. Lemay and Ralph E. Oesper, "M. E. Chevreul," *Journal of Chemical Education,* 1948, 25:66.

60 Dumas and Liebig, "Note sur l'état actuel de la chimie organique," *Comptes rendus de l'Académie des sciences,* 1837, 5:567–572, on p. 568.

as three different elements. Thus, at various times, Liebig, Wöhler, and Berzelius wrote benzoic acid as $[[(C^{14}H^{10}O^2) + O] + H^2O]$, where the benzoyl radical $(C^{14}H^{10}O^2)$ might be considered irreducible.

There was a duality within dualism and both approaches were consonant with Berzelius's principle of method.[61] The approach via the elements was the more common during the 1820s, but following the announcement of the benzoyl radical by Liebig and Friedrich Wöhler in 1832, the approach via putative radicals began to gain the ascendancy. It is true that the two approaches were compatible wherever a radical was supposed to contain two elements alone, but that should not be allowed to conceal the distinction between them. Berzelius himself, having developed the concept of organic radicals in the context of the vegetable acids, was inclined to favor the approach via the radicals. This tendency may account for the enthusiasm with which, in 1832, he greeted the benzoyl group. Precisely because it contained three different elements, it helped to differentiate the one form of dualism from the other. "The radical of benzoic acid," Berzelius exclaimed with delight, "is the first example proved with certainty of a ternary body possessing the properties of an element."[62]

As a consequence of the alternative approaches, there was considerable controversy during the 1830s among chemists who could all have said that they were following Berzelius's guidelines. Even among those who unequivocally opted for the radical theory, there was dissension when it came to the choice of radicals. It is certainly not true, as has often been implied, that the dualistic rubric was too narrow and restrictive in the organic world. The dispute between Liebig and Dumas concerning the formulation of ether and alcohol reflects the productive tension between the two approaches and the ascendancy of the radical theory, in that Dumas had capitulated by 1837. Both his etherin theory and the ethyl theory of Liebig were supported by analogy with inorganic compounds, and both conceived alcohol to be a hydrate of ether, as the latter could so easily be obtained from the former by dehydration. The dispute was over the construction of ether. Was it the hydrate of a hydrocarbon base, as Dumas suggested $(C^4H^8 + H^2O)$, or was ether itself the base and an oxide of the hypothetical ethyl radical, as Liebig maintained $(C^4H^{10} + O)$? Ether, for Dumas, was a dualistic compound in the sense that its constituents were each composed of two *elements;* for Liebig it was dualistic in the alternative sense that if the *radical* (C^4H^{10}) were analogous to a metal, then the ether would be an analog of an inorganic base (M +

61 Full documentation of the examples cited here is given in Brooke, "Chlorine Substitution," p. 79.
62 O. Theodor Benfey, ed., *Classics in the Theory of Chemical Combination* (New York: Dover, 1963), p. 38.

O). The details of the debate need not detain us, but the consequence of it was a vast increase in information about the properties and derivatives of alcohol.

The fact that the dualistic approach was stimulating and not stultifying should be clear from the behavior of Berzelius himself, who was quite prepared to change his mind if he felt that certain grooves looked unpromising. His volte-face over the benzoyl radical is probably the best known example. Having first welcomed it as a ternary radical, announcing the dawn of a new day in vegetable chemistry, he was declaring within two years that "Sauerstoff in einem Radikale ist ein nonsens."[63] The point (on which both Liebig and Dumas had similar reservations) was whether a radical as ostensibly electropositive as benzoyl could contain an element as highly electronegative as oxygen. Berzelius had eventually decided not,[64] and so transcribed the benzoyl compounds in such a way that their common radical became $(C^{14}H^{10})$ and not $(C^{14}H^{10}O^2)$. Benzoic acid thus became an analog of manganic acid – cf. $(C^{14}H^{10})O^3$ and MnO^3 – and the former benzoyl radical became an analog of the peroxide MnO^2. By the time Liebig reviewed the benzoyl derivatives in his textbook of organic chemistry, he had to discuss three competing theories, all consistent with dualism. There was the original benzoyl theory, the Berzelius transcription, and the theory of Mitscherlich who, impressed by the decarboxylation of benzoic acid, preferred to regard it as a compound of hydrocarbon and carbonic acid.[65] As with the alcohols, so with the benzoyl compounds, the dualistic presuppositions were not unduly restrictive. The analogy principle of Berzelius did not force every organic compound along a single track. It stimulated invaluable discussion and promoted the development of organic chemistry at a time when any other method would have foundered.

It is sometimes held against Berzelius – it certainly was by his French critics Laurent and Gerhardt – that his electrochemical theory blighted the prospects for a rational organic chemistry by requiring that two preformed groups be identified in every molecule. Insofar as organic radicals were supposed to behave like elements, there certainly was a sense in which Berzelius was committed to a kind of preformation. But his position can be easily misunderstood and it frequently has been. One has to ask whether, even in mineral chemistry, he regarded the electropositive and electronegative components as juxtaposed and preformed in the molecule. The standard critique leveled at his position by the unitary theorists of the 1840s was that it was naïve to represent barium sulfate, for example, in the form $[BaO + SO^3]$ as if the two oxides, albeit in a neu-

63 Partington, *History*, Vol. IV, on p. 329.   64 Berzelius, *JB*, 1834, *13*:197.
65 Liebig, *Traité*, Vol. I, p. 280.

tralized state, existed side by side in the salt.[66] The alleged naïveté was underlined by pointing out that barium sulfate could equally well be prepared from $BaO^2 + SO^2$ and from $BaS + 2O^2$. In which case, why should any one dualistic formula be privileged? It is a moot point, however, whether Berzelius ever deserved such criticism. If he had been guilty of that kind of crude preformationism, he had certainly renounced it by 1837, a whole decade before Laurent and Gerhardt caricatured him. It may well be the Laurent-Gerhardt critique that is responsible for much of the misapprehension on this point, for the French chemists assumed that one could not be a dualist without also being committed to a real preformation of the two component parts.[67]

What had Berzelius said on the subject? In a letter to Robert Hare in Philadelphia, he stated categorically that it was too crude to represent potassium sulfate by the one formula $[KO + SO^3]$. The alternative $[K + SO^4]$ must also be considered.[68] He evidently did not regard the salt as *containing* the two preformed groups KO and SO^3. Yet it is precisely that view, renounced by Berzelius for its crudity, which is still imputed to him.[69] This would be excusable if it was only in his private correspondence that he renounced the preformationist interpretation, but the fact is that he renounced it publicly all the time. In the *Jahresberichte* for 1833 and 1834 he confirmed[70] that stannous selenate could be represented as well by $[Sn + SeO^4]$ and $[SnSe + 4O]$ as by $[SnO + SeO^3]$. A year later he insisted that if he could but observe the individual atoms in copper sulfate we should discern neither CuO nor SO^3, but only one "single coherent substance."[71] If he could say this of inorganic compounds, it is difficult to see why he should be accused of having vitiated the understanding of organic compounds with a rigid dualism. It would seem more appropriate to say that he never entirely abandoned that view which he had expressed in 1814 – that organic compounds were, in a critical sense, unitary structures, in which the elements defied simple par-

---

66 Charles F. Gerhardt, *Introduction à l'etude de la chimie par le système unitaire* (Paris, 1848), p. 52.

67 The respective (and contrasting) moves made by Laurent and Gerhardt in differentiating their methodologies from those of their contemporaries are discussed in John H. Brooke, "Laurent, Gerhardt and the Philosophy of Chemistry," *Historical Studies in the Physical Sciences*, 1975, 6:405–429.

68 Berzelius to Robert Hare, 23 September 1834, in *Bref*, Vol. III:ii, pp. 141–142. Having discussed the two possible formulas, Berzelius concluded that "une théorie chimique plus développée nous montre de plus en plus la necessité de les considérer également toutes les deux."

69 Cf. Satish C. Kapoor, "The Origins of Laurent's Organic Classification," *Isis*, 1969, 60:477–527, on pp. 493–494.

70 Berzelius, *JB*, 1834, 13:185–188.   71 Ibid., 1835, 14:348.

titioning. And one can say this with some confidence because there was a sense in which it was true of inorganic combination also.

As noted in an earlier discussion on this point,[72] there was a distinguished chemist of the nineteenth century who once wrote that "so long as the simple atoms remain together," any one of the formulas [$CuS + O^4$], [$CuO^2 + SO^2$], [$Cu + SO^4$], [$CuO + SO^3$] was "as good as the other." But that distinguished chemist was not Gerhardt; it was none other than Berzelius himself. Chemical combination involved a rearrangement of elements, making it impossible to subscribe to a crude preformationism. As Eduard Farber observed, if Berzelius has to be accused of anything, it must be his dogmatic rejection, not a dogmatic adoption, of preformation.[73] Explaining his position to Wöhler, he pointed out that "one must know into which parts [an organic compound] can be divided, but it is not correct to say it is composed of these parts."[74]

The extension of dualistic precepts into organic chemistry was therefore neither implausible, nor unduly restrictive, nor crudely executed. If it became fashionable during the 1830s, it was largely because it proved to be compatible with several categories of empirical data. At the most rudimentary level, the very existence of organic acids and bases offered some support, as it was customary to classify all inorganic compounds as acid, base, or salt. Long after the substitution controversy had subsided, August W. Hofmann could write that it was *obvious* that a binary plan of combination obtained in a great many organic substances, the neutralization of benzoic acid with aniline (an organic base) providing a salient example.[75]

A stronger vindication of Berzelius's method came from the stoichiometric analogies that were drawn between the acids and bases of both domains. Berzelius[76] himself showed that if stearic acid were $C^{70}H^{134}O^5$, as analysis allowed, and if margaric acid were $C^{35}H^{67}O^3$, then they could be written in the form ($2R + 5O$) and ($R + 3O$), where $R = C^{35}H^{67}$. More impressive still was the identification of radicals that were common not just to two compounds but appeared to persist throughout a whole series of interconvertible substances. In this way the putative analogy between organic radical and inorganic element was reinforced. That is, of course, why the work of Liebig and Wöhler on the benzoyl radical was

72 Brooke, "Chlorine Substitution," p. 82.
73 Edward Farber, "Variants of Preformation Theory in the History of Chemistry," *Isis*, 1963, 54:443–460, on p. 455.
74 Berzelius to Friedrich Wöhler, 6 October 1834, cited by Farber, "Variants," on p. 454.
75 August W. Hofmann, "Royal Institution Lectures on Organic Chemistry," *Medical Times and Gazette*, 1853, 6:418.
76 Berzelius, *Traité*, Vol. V (Paris, 1831), p. 352.

so widely acclaimed. They had themselves concluded that their facts could be "arranged around a common centre, a group of atoms ($C^{14}H^{10}O^2$) preserving intact its nature, amid the most varied associations with other elements. This stability, this analogy, pervading all the phenomena, has induced us to consider this group as a sort of compound element."[77]

If not only the radical but peculiar physical properties also persisted throughout a series of organic compounds, it seemed even more reasonable to attribute the common properties to the presence of the common radical. Dumas adopted that line of reasoning in his *Leçons sur la philosophie chimique* of 1837. His predilection for preformed groups was sustained by Biot's discovery that the property of turpentine to deviate polarized light to the left was retained when the turpentine united with hydrochloric acid to form a new compound. A common radical and a common property survived together, and Dumas was much impressed.[78]

A yet more conclusive vindication was afforded by the isolation of organic radicals. If hypothetical radicals could be shown to enjoy an independent existence, their analogy with inorganic elements would be the more complete. This Gay-Lussac had ostensibly done as early as 1815 when he prepared cyanogen and compared it with the halogens.[79] The psychological impact of that one discovery was both considerable and enduring. Writing some 30 years later, Liebig could still say that "it is scarcely possible to imagine anything more wonderful than that carbon and nitrogen should form a gaseous compound, . . . which, in its properties and deportment, is a simple substance."[80] If one radical had been isolated why should not others follow?[81] In the mid-1830s Mitscherlich looked forward to the isolation of benzoyl on the supposition of its analogy with cyanogen.[82]

Finally, the analogy principle would have its choicest corroboration if a conjectured radical were both isolated and shown to exhibit a marked electrical character. The analogy with metals or halogens would then be as complete as possible, permitting a fusion of the two branches of chemistry. Such compelling evidence was in short supply, but it must not be forgotten that Robert Wilhelm Bunsen isolated what he thought was the cacodyl radical at the very time the substitutionists were mounting their

77 Benfey, *Classics*, on p. 15.
78 Dumas, *Leçons sur la philosophie chimique* (Paris, 1837), Lesson 9, pp. 348–349.
79 Crosland, *Gay-Lussac*, pp. 129–131.
80 Liebig, *Chemical Letters*, 1st ed., ed. John Gardner (London, 1843–1944), 2nd series, p. 216.
81 However, as Crosland points out (*Gay-Lussac*, p. 131), there was the complication that cyanogen, as a compound of only two elements, was initially regarded as inorganic; that is, it assumed its canonical status precisely when the pressure to isolate organic radicals began to mount.
82 Mitscherlich, *Élémens*, Vol. I, p. 119.

attack.[83] Bunsen claimed that his reaction between cacodyl chloride and zinc, in an atmosphere of carbon dioxide, had liberated the cacodyl radical – had shown that a hitherto hypothetical radical could be isolated in a free state. In ignorance of the fact that his radical had dimerized, he pushed the element/radical analogy as far as it would go. The cacodyl radical, Bunsen announced, "shares with the simple metals, and does so in an identical manner, the property of being able to combine directly with other substances."[84] There is a real sense in which Bunsen's classic research marked the culmination of the evidence in favor of the inorganic analogies dictated by Berzelius's principles. This was clearly recognized by those who continued to defend a dualistic approach. Edward Frankland, for example, spoke of Bunsen's work as a "most remarkable confirmation of the theory of organic radicals, as propounded by Berzelius and Liebig."[85] There is irony in the fact that the roof was placed on the dualist edifice at precisely the moment when its foundations began to shake, but it was that very coincidence that explains why the debates concerning the implications of chlorine substitution were so heated. Scarcely was the analogy principle of Berzelius more plausible than at the moment of its subversion.[86]

Before turning to the substitution controversy, which did so much to tarnish Berzelius's reputation, it is important to reflect on the advances made during the 1830s on the basis of the dualistic program. The fact that it was eventually transcended should not keep us from seeing that it appeared to be proving its worth in terms of solid achievement, so much so that Berzelius was loath to relinquish it. A previous analysis suggested there were at least five respects in which organic chemistry gained momentum from the dualistic analogies,[87] and they are summarized here.

The construction of analogies between organic and inorganic compounds, on the basis of Berzelius's dictum, generated rather than stifled controversy. If the debate between Liebig and Dumas over the formulation of ether and alcohol was infused with personal and national rivalry, it was also concerned with the choice of the most apposite analogs to be

83 Partington, *History*, Vol. IV, pp. 283–286.

84 Robert W. Bunsen, "Mémoire sur l'acide cacodylique et sur le sulfure de cacodyle," *Annales de chimie*, 1843, 8:356–362 on pp. 356–357.

85 Edward Frankland, "On a New Series of Organic Bodies Containing Metals," *Philosophical Transactions*, 1852, 142:417–444.

86 This point was, of course, perfectly clear to many nineteenth-century commentators. Thus Adolph von Baeyer could say of Bunsen's work that "the discovery of a compound organic metal, which inflamed in air and showed similar properties to potassium and sodium, removed the last doubt among the adherents of the radical theory of the correctness of Berzelius's teaching, that the organic world is a reproduction of the inorganic." Cited by Partington, *History*, Vol. IV, on p. 285.

87 Brooke, "Chlorine Substitution," pp. 87–90.

drawn from the inorganic repertoire. As a consequence of that debate, however misconceived it may seem to modern eyes, a wealth of information about the derivatives of alcohol accumulated during the 1830s – to such an extent that the two protagonists defied anyone to doubt that their dispute had stimulated fine and useful research.

More tangibly, perhaps, the application of Berzelius's method led to the identification of a number of functional groups of elements, an understanding of which has been germane to the development of organic chemistry. It is true that many of the radicals claimed for the organic acids proved to be mythical, but this was not true of groups such as ethyl, methyl, benzoyl, amide, and their relatives. Though the rationalization of substitution reactions was to challenge the electrochemical foundations of Berzelius's dualism, the concept of organic radicals continued to be productive. After he had renounced an electrical mechanism for organic combination, Dumas could still be predicting the imminent isolation of the ethyl radical.[88] A less sympathetic account would stress that claims such as those of Bunsen, Kolbe, and Frankland to have isolated hypothetical radicals rested on an illusion that was fed by the current inability to establish definitive molecular weights. That can hardly be denied; but the search for organic radicals and the attempts to characterize them were not themselves misconceived.

In suggesting a more positive evaluation of Berzelius's method, it must be remembered that it was self-consciously adopted by Liebig and Dumas in 1837 as the foundation of a joint research program, the object of which was to establish the identity of organic radicals from the reactions of their compounds. Liebig's work on uric acid, which conformed to that program, shows how productive it could be. As a prelude to his joint paper with Wöhler on uric acid and its derivatives, Liebig set the scene with an argument that illustrates the substantive as well as rhetorical function of method discourse:

> If we admit the principle that no ternary or quarternary compound can be formed except by the union of a binary compound with an element, or of two binary compounds with one another, it is clear that any further investigation of uric acid must be carried on with the intention of discovering the compound elements into which it may be resolved.[89]

88 Dumas, "Note . . . au sujet du mémoire de M. Bunsen," *Annales de Chimie*, 1843, 8:362–363.

89 Liebig, "On the Products of the Decomposition of Uric Acid," *Notices and Abstracts of Communications to the British Association for the Advancement of Science*, September 1837, p. 39.

No better testimony to the productivity of that principle could be adduced than that, when Liebig and Wöhler undertook their investigation, their research yielded no fewer than 16 new compounds, almost all of which were of major biochemical significance.[90] Even Liebig's adversary, his former pupil Gerhardt, was later to concede that the radical theory had indeed inspired remarkable work.[91]

Advances associated with the radical theory, however, by no means exhausted the potential of the analogy principle. There were other concepts that, when transferred from inorganic to organic chemistry became peculiarly effective. One such was the polybasicity of acids. Visiting Britain in 1837, Liebig came into contact with Thomas Graham, whose work on the acids of phosphorus was to earn him great acclaim. It was a serious oversimplification, Graham argued, to suppose that all acids could combine with only one proportion of base. The acids of phosphorus, for example, could be assigned different basicities according to their water content. Thus phosphoric became tribasic [P^2O^5 + 3 aq.]; pyrophosphoric dibasic [P^2O^5 + 2 aq.]; and metaphosphoric monobasic [P^2O^5 + aq.]. Liebig was evidently impressed by Graham's ideas, and in 1838 he published a paper on the constitution of the *organic* acids in which he endeavored to apply the new concept to them. After discussing the analysis of nine organic acids, he concluded that there were many in which a "perfect analogy" with the acids of phosphorus and arsenic would obtain.[92] It has to be said that Berzelius himself had not been won over to this particular innovation; but the fact remains that his analogy principle, in the hands of another, led to a crucial expansion of the concept of acidity.[93] For Liebig, it was a matter of some moment that the tartaric and phosphoric acids had polybasicity in common.[94]

Finally, the projection of dualism into organic chemistry had the effect of emphasizing the similarities between organic and inorganic compounds at the expense of their differences, thereby minimizing the need for a vital force or a vital principle. In a sense this was the working out of Berzelius's earlier predisposition in favor of a unified chemistry and against the creation of a separate ontology for the rationalization of or-

---

90 August W. Hofmann, *The Life Work of Liebig* (London, 1876), p. 85.
91 Gerhardt, *Précis de chimie organique*, 2 vols. (Paris, 1844–1845), Vol. I, p. 6.
92 Liebig, "Sur la constitution des acides organiques," *Annales de chimie*, 1838, 68:5–93, on p. 42.
93 For a fuller account of the significance of Graham's work in its relation to Liebig's reconsideration of the oxacid/hydracid distinction, see Frederic L. Holmes, "Liebig," in Charles C. Gillispie, ed., *Dictionary of Scientific Biography*, Vol. VIII (New York and London, 1973), pp. 329–350, on pp. 338–341.
94 Liebig, "Acides organiques," p. 55.

ganic substances. As a consequence of the application of Berzelius's analogy principle during the 1830s, sufficient evidence accumulated to support the contention that organic compounds did essentially resemble those of the inorganic world. It was in their joint statement of 1837 that Liebig and Dumas boldly stated that the *only* difference between organic and inorganic chemistry was that in organic chemistry the radicals were complex.[95] More than the artificial synthesis of organic compounds, the systematic attempt to assimilate them to those of the mineral realm, via the categories of dualism, helped to erode the possibility of a vitalism erected on their differences.[96] It was one of Berzelius's detractors, Gerhardt, who still felt the need of a vital force in 1842, though even he soon transcribed it into a "reducing force."[97] In addition to spawning fruitful controversy, in addition to focusing attention on the fertile concept of organic radicals, in addition to underwriting the research programs of others, and in addition to justifying the transfer of concepts such as polybasicity from one domain to the other, Berzelius's method also had the merit of tacitly excluding vital agents from the context of chemical practice, even if they continued to find a niche in the description of physiological functions. These were some of the gains during the 1830s. They were not, however, sufficient to prevent on all-out attack on Berzelius's chemistry as that decade came to a close. The form of the attack, and of Berzelius's reaction to it, must now be considered.

Berzelius and the Controversy over Chlorine Substitution

As early as 1833, Dumas was conducting experiments on the chlorination of hydrocarbons that he extended to other organic products, such as alcohol. One of his conclusions was that chlorine had the singular power of separating hydrogen from certain bodies and of replacing it atom for atom. In an attempt to codify his results in quantitative terms, he had announced a series of rules, the first of which read as follows: "When a substance containing hydrogen is submitted to the dehydrogenating action of chlorine, of bromine, of iodine, of oxygen, etc., for each atom of hydrogen which it loses it gains an atom of chlorine, of bromine, of iodine, or half an atom of oxygen."[98] A different rule applied if the organic

95 Dumas and Liebig, "L'état actuel," pp. 567–572.
96 Brooke, "Organic Synthesis."
97 Gerhardt, "Recherches sur la classification chimique des substances organiques," *Revue scientifique et industrielle de Quesneville*, 1842, *10*:145–218, and 1843, *14*:592–600.
98 Dumas, "Considérations générales sur la composition théorique des matières organiques, troisième partie," *Journal de Pharmacie*, 1834, *20*:261–294, on pp. 285–286. For a recent account of Dumas's research programs during the 1830s, see Leo J. Klos-

compound contained water, as Dumas at the time believed alcohol to do. In such empirical generalizations there was no particular threat to Berzelius's chemistry.

Events took a more serious turn, however, when Dumas's assistant, Laurent, studied the substitution products of naphthalene for his doctoral thesis.[99] For out of that work came the more radical proposal that in certain organic compounds chlorine not only displaced hydrogen but played its role, so little were the properties affected. But if an electronegative element, chlorine, could play the role of an electropositive element, hydrogen, the implications were startling. To make an electrochemical theory the basis for rationalizing chemical reactions, whether in the organic or inorganic sphere, could begin to look hopelessly misguided.[100] The challenge to Berzelius's authority soon took the form of theories of "types," which ascribed the properties of a compound more to the arrangement of the elements within it than to their electrical character. In Laurent's doctoral thesis and throughout his subsequent career, models of atomic arrangement were even projected into three dimensions, by analogy with crystallographic structures.[101] It is well known that Berzelius was not amused. The propositions through which Laurent presented his thesis were condemned as "a kind of legislation in organic chemistry which, like all hasty legislation, appears in numerous regulations." With even greater condescension, Berzelius decided that it would be useless for his reports to deal in the future with such theories.[102]

Despite the several excellent accounts of the ensuing controversy,[103] we still lack the kind of in-depth analysis that Martin Rudwick produced

terman, "A Research School of Chemistry in the Nineteenth Century: Jean Baptiste Dumas and His Research Students," *Annals of Science*, 1985, 42:1–80, especially pp. 49–56.

99 Auguste Laurent, "Théorie des combinaisons organiques," *Annales de chimie*, 1836, 61:125–146. See also Jean Jacques, "La thèse de doctorat d'Auguste Laurent et la théorie des combinaisons organiques (1836)," *Bulletin société chimique*, May 1954, suppl. 31–39.

100 This is not to say that Laurent's nucleus theory was *conceived* as a way of attacking electrochemical dualism. In his doctoral thesis he wrote that it was his intention to reconcile the constitutional goals of dualism with the theoretical considerations that now seemed to frustrate them; Brooke, "Philosophy of Chemistry," p. 408.

101 Kapoor, "Laurent's Organic Classification."

102 Partington, *History*, Vol. IV, p. 381.

103 For useful discussion, in addition to Kapoor (n. 69), Klosterman (n. 98), and Levere (n. 3), of the relationship between radical and type theories see Nicholas W. Fisher, "Organic Classification before Kekulé," *Ambix*, 1973, 20:106–131, 209–233; Alan J. Rocke, *Chemical Atomism in the Nineteenth Century* (Columbus: Ohio State University, 1984); and Colin A. Russell, *The History of Valency* (Leicester: Leicester University Press, 1971), Chs. 2 and 3.

for a great geological controversy that was taking place almost concurrently.[104] Rudwick's analysis of the Devonian controversy is, however, invaluable for specifying the desiderata that a polished analysis would have to meet. There are questions to be asked about the trajectory of each participant, about the extent to which positions were modified as a result of new data and as a result of the dialectics of controversy. There is the question whether relatively minor figures played a role in producing these data. And there is the question of how the perceived competence of each of the participants affected the credibility and the weight of their testimony. And if a consensus was finally achieved, was it one of the initial positions that was vindicated, or did neither side, strictly speaking, win?[105]

This essay is yet another that can only scratch the surface of what was one of the most colorful and in some ways distressing controversies in the entire history of chemistry. It is quite clear, however, that the questions Rudwick has posed would be just as fruitful in this chemical as in his geological context. The trajectories of some of the participants deserve the closest attention, not merely to track the course of the debate but because changes in position were charged with significance in the context of priority claims. The most famous example would be the volte-face of Dumas, who, having first dissociated himself from Laurent's claim that chlorine could play the role of hydrogen, subsequently incensed his former protégé by incorporating that very idea into a theory of types that Dumas claimed as his own. From Dumas's standpoint, he had simply refused to be persuaded until more evidence than Laurent could muster had been forthcoming, notably his own work on the chlorination of acetic acid, which he seems to have regarded as a test case.[106] But from Laurent's standpoint, the dishonesty was flagrant, the imputation that his own work had been hasty was hurtful, and, as one consigned to the provinces, the seemingly irreparable damage to his career meant that he harbored a grudge for years to come − a grudge that even found its way into the titles of his papers.[107]

There is no doubt, too, that positions were modified in response to new experimental work. If Dumas's own testimony is to be believed, his initial reluctance to abandon a dualistic paradigm was overcome as a consequence of the work of F. J. M. Malaguti on the chlorination of

104 Martin J. S. Rudwick, *The Great Devonian Controversy: The Shaping of Scientific Knowledge among Gentlemanly Specialists* (Chicago: Chicago University Press, 1985).
105 Ibid., part 3.   106 Klosterman, "Research School," pp. 53–54.
107 In several of his memoirs on chemical types, Laurent would insert in the title that they were types "not discovered by M. Dumas." The fullest documentation of the grievances of both Laurent and Gerhardt remains L. E. Grimaux and Charles Gerhardt (fils), *Charles Gerhardt: Sa vie, son oeuvre, sa correspondance* (Paris, 1900).

ethers, the work of Henri Victor Regnault on Dutch oil, and finally his own success, in August 1838, in preparing a pure chloracetic acid in which the essential properties of the parent acid were retained.[108] The experimental work of relatively minor figures must also be taken into account. It has long been recognized that the apotheosis of substitution theory came when another of Dumas's assistants, L. H. F. Melsens, found that trichloracetic acid could be converted back to acetic by the use of hydrogen generated by the action of acid on potassium amalgam.[109] The facility with which substitution and inverse substitution could be achieved meant that by 1844 there was an effective consensus: Acetic and tri-chloracetic acids must have a similar formula, must belong to the same chemical type.

In his discussion of the gradient of attributed competence, Rudwick notes that any map of the topography must allow for the movement of individuals across it: "Scientific status, like the stock market, could go down as well as up."[110] One reason why the substitution controversy merits further investigation is that it shows how the weight of attributed competence could change as a consequence of the stance adopted by participants on issues endemic in the controversy. There is no doubt that the authority Berzelius had wielded through the *Jahresberichte* was drained as a consequence of the contortions he went through to save face. In May 1838, before he had switched allegiance, Dumas was still prepared to be deferential. Indeed, it was in the knowledge that he was on the point of losing esteem in Berzelius's eyes that he swiftly dissociated himself from Laurent's theory. In a letter to Théophile Jules Pelouze, Berzelius had expressed his regret that Dumas was advocating the idea that a chlorine atom could replace a hydrogen atom, a notion that appeared to lie behind Laurent's "complicated and bizarre views," a notion, in short, that was contrary to the first principles of chemistry.[111] In seeking to put the record straight, Dumas committed himself to a form of words that, while they might have appeased Berzelius, became the Achilles heel in his relationship with Laurent. Disavowing the hypothesis that chlorine could take the *place* of hydrogen (let alone play its role), he insisted that he was "not responsible for the gross exaggeration with which Laurent has invested my theory; his analyses moreover do not merit any confidence."[112]

108 Kosterman, "Research School," p. 53.

109 Ibid., p. 55; Partington, *History*, Vol. IV, p. 364.

110 Rudwick, *Great Devonian Controversy*, p. 420.

111 These sentiments of Berzelius were made public in "Lettre de M. Berzelius à M. Pelouze," *Comptes rendus de l'Académie des Sciences*, 1838, 6:629–644; a "Notes de M. Pelouze" followed on pp. 644–645.

112 Dumas, "Observations sur les communications précédentes [de M. Berzelius et M. Pelouze],"; ibid., 1838, 6:645–648.

One can understand Laurent's chagrin when Dumas's trajectory landed him in a position that looked like a gross exaggeration of *his* own – Dumas eventually suggesting that even carbon might be substituted by chlorine in certain organic compounds! Evidently by 1840, when Dumas allowed himself that indulgence,[113] deference toward Berzelius was no longer on the agenda. And the same was true for Liebig, who, in a letter to Wöhler, stated that Berzelius was fighting for a lost cause, that he would much better leave things to those who still had something to achieve.[114] The visual evidence that persuaded contemporaries that Berzelius was no longer at the rock face was the sheer complexity of the formulas he initially devised to prevent hydrogen and chlorine from occupying the same place in the molecule.[115]

Liebig's reference to a lost cause raises the question whether it was a simple case of victory for the one side. Was there a relatively swift paradigm shift in which the electrical nature of the elements gave way to their physical arrangement as the principal determinant of chemical properties? Or were concessions made by both sides as the data base was broadened in scope? As with Rudwick's Devonian controversy, the latter is much nearer the mark. By 1843, when Berzelius was in correspondence with Laurent, he had in effect conceded the point that chlorine could occupy the same place in the molecule as hydrogen.[116] For his part, Laurent, too, had moderated his position: "Certainly," he acknowledged, "the chlorine introduces modifications into these compounds."[117] These two concessions having been made, it was a puzzle to Laurent to know why they need be at loggerheads at all. "I think," he ventured to suggest, "you have granted me all I ask."[118] To the French chemist it finally seemed that the entire debate had fizzled out into a verbal quibble: "Whether we say that the chlorine plays or does not play the role of hydrogen is of little import."

The fact that Berzelius was eventually prepared to make some concession may suggest that he was perhaps not quite the obscurantist that historians have commonly made him. In the remainder of this section it is argued that it is worthwhile to try to understand Berzelius's position,

113 Dumas, "Mémoire sur la loi des substitutions et la théorie des types," *Comptes rendus de l'Académie des Sciences*,1840, 10:149–178, on p. 155: "on peut faire subir de véritables substitutions au carbone. . . ."
114 Partington, *History,* Vol. IV, p. 370.
115 Examples of these cumbersome formulas are given in ibid., pp. 368–369. Chlorinated ethyl formate, for example, looked like this: $[(2C^2H^2O^3 + C^2H^2Cl^6) + (2C^4H^6O^3 + C^4H^6Cl^6)]$.
116 Berzelius to Laurent, October 1843, in *Bref,* Vol. III:ii, p. 188; Brooke, "Chlorine Substitution," p. 56.
117 Laurent to Berzelius, May 1843, *Bref,* Vol. III:ii, p. 184.
118 Laurent to Berzelius, January 1844, ibid., p. 194.

for a sympathetic reassessment is possible. It was important that the more extravagant notions of the substitutionists should be subjected to criticism. In several respects their case was not self-evidently correct, and it was in some measure due to Berzelius that a greater rigor was introduced into the discussion. The basic contention that chlorine could simulate the function of hydrogen in an organic compound had been open to several objections, four of which carried most of the weight.

First, one could acknowledge that a parent compound and its chlorine derivative did resemble each other in a few trivial respects, but insist that their characteristic properties differed. This objection revolved around the additional problem of deciding what should, or should not, constitute characteristic properties. When Dumas stressed the parallel behavior of acetic and trichloroacetic acids during alkaline decomposition (the former yielding methane, the latter chloroform), Pelouze, who tended to identify with Berzelius, simply retaliated that these reactions were merely two instances of a property shared by all compounds of carbon, hydrogen, and oxygen. Again, if Dumas pinpointed *chemical* analogies between acetic and chloroacetic acids, Berzelius would retaliate by pointing to disparate *physical* properties. Because the products of organic reactions depend so largely on the conditions employed, it was usually possible, granted sufficient ingenuity, to find convincing differences between two supposedly analogous compounds, as when Berzelius insisted on the dissimilar behavior of potassium acetate and chloroacetate when treated with potassium hydroxide, the former producing acetone by eventual dry distillation, the latter giving rise to chloroform.[119]

Second, proponents of chlorine substitution were indeed prone to overgeneralization which openly invited contradiction. One of the reasons why Laurent was so annoyed when Dumas chanced his arm with the possibility of carbon substitution was that such excesses were likely to damage the chances of the substitutionist cause. Dumas duly had his comeuppance when Wöhler published his satirical note in Liebig's *Annalen,*[120] in which it was reported that he, S. C. H. Windler, had replaced all the elements of manganese acetate with chlorine, the wonderful product enjoying the formula Cl^{24}. In other respects the once cautious Dumas was now going further than Laurent, for all his boldness, would countenance. The younger chemist had been much more apprehensive about the mutual displacement of hydrogen and oxygen than his senior,[121] and he could also protest that Dumas's earliest views on the nature of substitu-

119 Berzelius, "Sur la théorie les substitutions et des types chimiques de M. Dumas," *Journal de pharmacie*, 1840, 26:597–614, on p. 612.

120 S. C. H. Windler [Wöhler], "Ueber das Substitutionsgesetz und die Theorie der Typen," *Annalen der Chemie*, 1843, 33:308–310.

121 Laurent, *Chemical Method*, tr. William Odling (London, 1855), p. 299.

tion reactions had precluded the recognition that there were whole series of chlorination reactions where equivalent substitution could not possibly be involved – reactions such as the *addition* of chlorine to the naphthalenes.[122] Moreover, when Dumas had eventually enunciated his type theory, according to which *every* substitution reaction proceeded without disrupting a primitive "type" or arrangement of elements, Laurent had to dissociate himself yet again. To provide for those substitution reactions that resulted in a palpable modification of the original chemical properties, Dumas had devised what he called "mechanical types."[123] This device, so Laurent informed Berzelius,[124] was entirely ad hoc. We therefore have the engaging spectacle of Berzelius correcting Dumas with arguments that Laurent had provided.[125] Far from being obscurantist, he was quite justified in saying of Dumas's conclusions that they were true in some cases and erroneous in others.[126]

A third objection centered on the problem of proof. Familiar with a whole series of chlorinated naphthalenes and derivatives of isatin, Laurent was to say that "color, solubility, volatility, vapor density, crystalline form . . . action upon polarized light, action upon the animal economy, all testified in favor of the theory of substitutions."[127] But an empiricist could always object that "testify in favor of" was not the same as proof. Though one might grant an impressive similarity between parent and derivative, this similarity did not entail the inference that the chlorine was occupying the exact place vacated by the hydrogen. Because one had no access to the interior of a molecule, one had no rigorous proof of exact displacement. Critics could make much of this lack.

That acetic and trichloroacetic acids could be converted into methane and chloroform, respectively, by no means convinced N. A. E. Millon and Pelouze that the two acids belonged to the same type. They could acknowledge the family resemblance between methane, CH^4, and chloroform, $CHCl^3$, without having to accept that the respective acids from which they were derived necessarily embodied an identical arrangement of their elements. In general terms, their objection was that two quite different "types," two quite different arrangements of elements, could engender analogous products by an appropriate, but in each case differ-

---

122 Laurent, "Sur les acides pimarique, pyromarique, azomarique, etc.," *Annales de chimie,* 1839, 72:383–415, on pp. 411–414.
123 Dumas, "Loi des substitutions," p. 162.
124 Laurent to Berzelius, 12 May 1843, in *Bref,* Vol. III:ii, p. 182.
125 See, for example, Berzelius's remarks in "Lettre de M. Berzelius à M. Pelouze," *Comptes rendus de l'Académie des sciences,* 1838, 6:629–644, on p. 634.
126 Berzelius, "Types chimiques de Dumas," p. 601.
127 Laurent, *Chemical Method,* pp. 202–203.

ent, scrambling of the respective primitive arrangements.[128] It was difficult to prove that analogous products had structurally analogous antecedents. Dumas confessed as much when he acknowledged that his concept of chemical and mechanical types transcended the empirical level. The idea of persistent patterns of internal arrangement went further than the facts,[129] and for that very reason the burden of proof remained with the innovators.

Their difficulties are nicely illustrated by a lapse into circularity that Laurent did not always manage to avoid. He would affirm in one breath that chlorine could only play the role of hydrogen when it occupied exactly the same place in the molecule. How could one know when such exact displacement had occurred? In the next breath he would reply that one knows it is so "when the properties of the chlorinated substance are analogous to those of the original."[130] It was almost as if knowledge of identical roles presupposed knowledge of exact replacement, while at the same time knowledge of exact replacement presupposed knowledge of more or less identical roles. As argued elsewhere, the circle was broken by the adoption of a metaphysical principle – that of minimal structural change – which is still employed today and sometimes as unconsciously as it was by Berzelius's critics.[131] In the last analysis, the substitutionists could not prove their case. A geometrically precise replacement of hydrogen by chlorine tended to be assumed, not demonstrated. Laurent could invoke simplicity and aesthetic criteria to vindicate his views, as he did when writing to Berzelius; but his correspondent, adamant and obstinate though he may have been, was strictly speaking correct when he insisted that the idea of arrangement types rested on inaccessible foundations.[132]

The fourth objection brings us full circle to the methodological principle that Berzelius reiterated in his reply to Laurent. The claim that chlorine played the role of hydrogen could always be resisted with the observation that even if parent and derivative did exhibit analogous properties, they were, after all, analogous and not identical. Trichloracetic acid, for example, is actually a stronger acid than acetic. The point for which Berzelius stuck out was that not all the genuine differences between elementary hydrogen and chlorine could be smothered in organic combination. And it was to inorganic combination he returned to drive home the message. He invited Laurent to consider the effect of chlorinating phosphine:

128 N. A. E. Millon and T. J. Pelouze, "Note sur la décomposition des substances organiques par la baryte," *Comptes rendus de l'Académie des sciences,* 1840, 10:48–50.
129 Dumas, "Loi des substitutions," p. 177.
130 Laurent to Berzelius, 12 May 1843, in *Bref,* Vol. III:ii, p. 182.
131 Brooke, "Chlorine Substitution," pp. 53–54.
132 Berzelius, "Types chimiques de Dumas," pp. 603–604.

The product, phosphorus trichloride, diverged considerably from the parent compound, as the chlorine made its mark on the molecule. It was inconceivable, in either the organic or inorganic domain, that no such mark should be made.[133]

Looking to the future of chemistry, Berzelius was alarmed by the manner in which Laurent was inverting his principle of method, taking his own idiosyncratic theory of organic compounds as the basis for revising inorganic chemistry, the central concepts of which had been secured until then by electrochemical theory. For Berzelius there was a certain lack of proportion if one organic reaction could be used to topple the entire theoretical edifice he had so carefully constructed. While it would be incorrect to imply that Laurent's nucleus theory arose sui generis from the study of that one kind of reaction – it was deeply indebted to the crystallographic models of Haüy[134] – Berzelius was nevertheless correct in asserting that the conventions of electrochemical dualism were being subverted as a consequence of analogies projected from the organic realm. Nor did Laurent conceal the fact. He explicitly announced his intention of extending into inorganic chemistry any inferences he might draw from his study of the naphthalene compounds.[135] The following examples, distilled from Laurent's chemistry, are indicative of the consequences when Berzelius's method was turned upside down.

A striking reinterpretation of inorganic concepts occurred in the context of isomorphism. Both Laurent and Dumas wished to exploit an analogy between chlorine substitution in organic chemistry and the fact that two inorganic compounds like potassium permanganate and perchlorate could crystallize in the same form, despite the great differences between elementary chlorine and manganese. If chlorine could replace manganese in inorganic compounds without effect, why should chlorine not replace hydrogen in organic compounds without effect? It was the alums he quoted in his correspondence with Berzelius, but the point was the same: All the evidence supported the view that "in some cases chlorine plays the role of hydrogen, just as in the alums oxide of chromium plays that of alumina."[136] How did Berzelius react to this analogy? In fact, he ignored it, regarding it as too fragile to bear the weight put upon it. Like his former pupil, Mitscherlich, who had pioneered the study of isomorphism, Berzelius had always considered the retention of crystal form to be a purely *mechanical* effect, having nothing to do with chemical affinity, and there-

133 Berzelius to Laurent, 20 October 1843, in *Bref,* Vol. III:ii, p. 189.
134 Kapoor, "Laurent's Organic Classification."
135 See, for example, his remarks in his "Sur la série naphthalique, 31me mémoire," *Revue scientifique et industrielle de Quesnevile,* 1843, 14:74–113, 313–349, 556–580, on p. 74.
136 Laurent to Berzelius, 12 May 1843, in *Bref,* Vol. III:ii, p. 181.

fore beyond the jurisdiction of electrochemical theory. Because he remained committed to the view that isomorphism was "only a mechanical
phenomenon of crystallization,"[137] Berzelius remained aloof from Dumas's claim that it was actually in "very little agreement with the electrochemical theory."[138] The fact is that the analogy exploited by Laurent
and Dumas carried plausibility only if the phenomenon of isomorphism
had already been reinterpreted in the light of substitution theory.[139]

A second example relates to the definition of organic radicals. During
the 1830s the concept had been defined by Liebig and Dumas, as well as
by Berzelius, in accordance with an analogy derived from the inorganic
elements. "By compound radicals or organic radicals," Liebig explained,
"we mean certain complex bodies which have the property of forming
compounds with the elements which are analogous to those formed by
the elements among themselves."[140] Groups like benzoyl, cyanogen, ammonium, and ethyl had been baptized as radicals precisely because they
persisted throughout a series of compounds as did the elements themselves. Whatever the origins of Laurent's concept of hydrocarbon radicals, it is particularly significant that his definition no longer required
reference to inorganic analogs. He could define a radical as a neutral,
stable hydrocarbon within which substitutions might occur. This transition could not have escaped Berzelius's notice, for he made a point of
reporting that Dumas likewise was no longer comparing organic radicals
with the simple elements.[141]

Laurent was, in fact, doing the very reverse. He began to compare the
supposedly simple elements with his complex hydrocarbon radicals. Inverting Berzelius's method yet again, Laurent was one of the first to argue
from the complexity of his organic radicals to the complexity of the elements. If different hydrocarbon radicals could be built from different
degrees of condensation of ulterior CH units, why should the elements
not be compounded from different degrees of condensation of their ulterior units?[142] Laurent was attracted to such a possibility because it
would help to reconcile the electrochemical and substitution theories:
Perhaps one and the same element, such as chlorine, could appear in
different electrical guises depending on the arrangement of ulterior but

137 Berzelius, *Traité*, Vol. VI, 2nd ed. (Paris, 1850), p. 752.
138 Dumas, "Mémoire sur la constitution de quelques corps organiques et sur la théorie
des substitutions," *Comptes rendus de l'Académie des sciences*, 1839, 8:609–622, on
p. 609.
139 This point is argued in greater depth in Brooke, "Chlorine Substitution," p. 69.
140 Liebig, *Traité*, Vol. I, p. 1.
141 Berzelius, "Opinions Relating to the Composition of Organic Substances," *Taylor's
Scientific Memoirs*, 1846, 4:668.
142 Laurent, "Sur la série naphthalique," p. 100.

identical components? But such conjectures were simply too speculative for Berzelius, who dug in his heels on the ground that the question of the complexity of the elements should be shelved until there were analytical data in favor.[143]

The reformulation of acids provides a third example of the inversion of Berzelius's method. At just the time Laurent engaged in correspondence with Berzelius, he was toying with a distinctively new model. Contrary to previous opinion, and contrary to what he himself would later believe, he chose for a while to regard an organic acid as a structure in which oxygen lay exterior to a hydrocarbon prism. Acetic acid he would represent by the formula $[C^8H^8 + O^4]$ and formic acid by $[C^4H^4 + O^4]$. This was not a wild conjecture, as Laurent felt he had confirmed the hypothesis when he succeeded in oxidizing the hydrocarbon stilbene to yield benzaldehyde and benzoic acid.[144] But it was a weak conjecture, as even his closest colleagues had to admit. Weak though it was, he went straight ahead and rewrote the inorganic acids in this new fashion, indicating again the strength of his commitment to a method that, while still solicitous of the unity of chemistry, was, as Berzelius properly realized, diametrically opposite to his own. Phosphoric acid became $[P^2H^6 + O^4]$, sulfuric $[SH + O^4]$, and in Laurent's own words, "and so on for the mineral salts."[145] To Berzelius, who had continued to uphold his Lavoisian heritage, such a stroke must have seemed rash in the extreme.[146]

Though the adequacy of the conventional anhydride-plus-water model had been questioned for the organic acids, and especially by Liebig, there was a fairly general resistance to the revision of the inorganic acids. Even Liebig himself, despite his testimony that all his classic research on the basicity of acids was inspired by the alternative precepts of Humphry Davy, stated that the alternative approach (in which displaceable *hydrogen* was stressed) had no place in inorganic chemistry.[147] It is not in the least surprising that Berzelius should have resisted the deviant view of Laurent that ascribed the origin of acidity, in inorganic as well as organic acids, neither to displaceable water, nor to displaceable hydrogen, but to the location of peripheral oxygen.

Berzelius was substantially correct in his assessment that Laurent and

---

143 For an introduction to nineteenth-century literature on the complexity of the elements, see David M. Knight, *The Transcendental Part of Chemistry* (Folkestone: Dawson, 1978).

144 Laurent, "Nouvelles recherches sur la théorie des radicaux dérivées," *Comptes rendus de L'Académie des sciences*, 1843, 16:340–343, on p. 341.

145 Laurent, "Sur la série naphthalique," p. 110 [my translation].

146 On the relationship between the chemistry of Lavoisier and that of Berzelius, see Melhado, *Berzelius*, especially pp. 159–164.

147 Liebig, "Acides organiques," p. 72.

other substitutionists were not merely developing alternative concepts to his own but plying a new method at variance with the one that had been well tried and tested. Had he lived longer, he would have witnessed an even more systematic inversion of his own approach. For in the chemistry of Gerhardt the doctrine of chlorine substitution became of such paramount importance that every single chemical reaction, every inorganic as well as organic process, became a substitution reaction. Berzelius's suspicions became incarnate in the unitary system of Gerhardt, which so methodically reinterpreted the whole fabric of inorganic chemistry that not a single addition reaction remained.[148] If the electrochemical approach to organic chemistry had erred by construing all combination as the addition of electropositive to electronegative constituents, so the unitary approach to inorganic chemistry erred equally by construing all reactions as substitutions. The simple addition of oxygen to potassium sulfide, when reinterpreted by Gerhardt, became a special case of double substitution in which the two products, potassium oxide and sulphur trioxide, for some reason declined to separate.[149] Taken to its logical conclusion, the method that Berzelius had so relentlessly opposed generated anomalies as extravagant as those imputed to his own.

Conclusion: Toward a Reassessment

I have tried to show in this discussion that the traditional assessment of Berzelius as one whose forays into organic chemistry were essentially misadventures is not well founded. It rests on an anachronistic approach to the chemistry of the 1830s, when it was far from clear that a dualistic scheme would prove abortive. It also overlooks the fact that Berzelius's approach did not prove abortive. It did much to sustain the durable concept of organic radicals and it generated fruitful controversy. Nor was Berzelius's stand on chlorine substitution groundless. It was not a simple matter of his being on the losing side, for he scored points against the French chemists when their claims became excessive and he did, if not immediately, concede the point that chlorine could directly substitute for hydrogen in an organic compound, occupying its place though never entirely replicating its role. A more sympathetic understanding of his predicament also becomes possible once the regulative character of his analogy principle is taken into account. He would not be seduced into revising

148 Gerhardt, *Système Unitaire*.
149 Instead of accepting a simple addition, $K^2S + 2O^2 \rightarrow K^2SO^4$, Gerhardt conceived the reaction thus: $K^2S + O^3 \cdot O \rightarrow [SO^3 + K^2O]$. An oxygen atom was, gratuitously, deemed to replace the sulfur; Gerhardt, *Traité de chimie organique*, 4 vols. (Paris, 1853–1856), Vol. IV, p. 571 et passim.

the conceptual basis of inorganic as well as organic chemistry on the strength of one anomaly in the organic sphere.

A balanced assessment must not, of course, gloss over the defects of his method, even if it took time for some of them to appear. The most familiar would be the constraint that electrochemical dualism placed on the possibility of polyatomic molecules having the form X^n. If all combination had to be explained by the neutralization of opposite electricities, then it was almost inconceivable that either two atoms, or two groups of atoms, of the same kind should combine or coalesce. Like would repel like, making any molecule of the form X^n completely unstable. This consideration was one of the more effective suppressors of Avogadro's hypothesis until its eventual triumph in the 1860s.[150] The extent to which a readier acceptance of that hypothesis would have clarified atomic and molecular weights during the 50 years following its enunciation has been exaggerated, but a rationalization of the kind proposed by Cannizzaro was certainly excluded by any form of electrochemical theory that precluded such diatomic molecules as H^2, Cl^2, and O^2.

While it would be absurd to blame Berzelius for the eclipse of Avogadro's insight, the failure to standardize chemical formulas created a situation in which seemingly propitious analogies between organic and inorganic species could be constructed, only to be dismantled later when molecular weights were revised. The failure of standardization is perhaps most visible in the case of the organic acids where the anhydride-plus-water model that Berzelius had nurtured eventually proved untenable. If, as Gerhardt insisted, the molecular formula of acetic acid had to be halved from $C^4H^8O^4$ to $C^2H^4O^2$, it was no longer possible to portray with consistency the organic acid as a simple analog of sulfuric. The comparison between $[SO^3 + H^2O]$ and $[C^4H^6 \cdot O^3 + H^2O]$ broke down. A full account of this particular problem would have to take into account the assumptions that Berzelius made when determining the organic acids from their silver salts, and the assumptions Gerhardt made in his case for standardizing organic with inorganic formulas; but there is no doubt that Berzelius's own application of his analogy principle did have the longer-term defect of encouraging a formulation of organic acids that, with hindsight, we can see was doomed.[151] For those such as Frankland and Kolbe, who

---

150 Recent studies of the fate of Avogadro's hypothesis include Nicholas W. Fisher, "Avogadro, the Chemists, and Historians of Chemistry," *History of Science*, 1982, 20:77–102, 212–231; and John H. Brooke, "Avogadro's Hypothesis and Its Fate: A Case-Study in the Failure of Case-Studies," *History of Science*, 1981, 19:235–273.

151 The gist of Gerhardt's revision was his insistence that organic and inorganic formulas should be determined by reference to the same number of volumes of vapor. Current practice was to speak of four-volume formulas for organic compounds and two-

continued to work within the confines of a dualistic mold, a similar and related problem arose, in that the isolation of the methyl and ethyl radicals that each thought he had achieved turned out to be chimerical once the molecular weights of the alkanes (which they had in fact prepared) proved that dimerization had occurred. That very dimerization (of two identical radicals) would have been inconceivable on dualistic principles where a strict application forbad the union of identical components.

A second defect that a balanced assessment would have to recognize was the constraint placed on both branches of chemistry by a framework in which all reactions had to be construed as either additions or separations – the addition of a positive to a negative group or the separation of two such groups from a stable compound. Substitution theory had the undeniable merit of introducing a different type of reaction, that of straightforward replacement. Reactions that, for Berzelius and his followers, had been conceived as a sequence of additions and separations could be simplified by reference to simultaneous and mutual displacement. Take the case of the chlorination of alcohol, which, according to one dualistic account, required the following stages: The chlorine first oxidizes alcohol to acetaldehyde; then the "carbonic oxide," supposedly preexisting within both alcohol and aldehyde, unites (i.e., adds to) the chlorine to give $CO^2 Cl^4$; the residual hydrocarbon unit C^2H^2 also absorbs chlorine to produce $(C^2H^2) Cl^2$; and, as the grand finale, these two intermediaries unite to generate the chloral $(C^2H^2Cl^2 + CO^2Cl^4)$. For the French chemist J. F. Persoz, writing in 1839, such a scheme was preferable to admitting the simple substitution of hydrogen by chlorine.[152]

So long as the plus symbol had to appear in every stage of every chemical reaction, the corresponding reaction schemes were apt to be cumbersome and restrictive. In the clarification of the relationship between alcohol and ether, eventually achieved by the English chemist Alexander Williamson, it was necessary to integrate concepts drawn from both radical and type theories in order to transcend the limitations of each. By relating the ether derived from alcohol to a water molecule in which each of the hydrogen atoms was replaced by an ethyl radical, Williamson emphasized that two molecules of alcohol were required to produce one of ether – a proposition he was able to corroborate by the preparation of

volume for inorganic. The inconsistency was particularly marked in the context of the organic acids, for the H^2O component in Berzelius's formulation was represented in its two-volume (inorganic) form, but *within* a formula that was determined by reference to four volumes. The aggressive manner in which Gerhardt sought to weed out this inconsistency meant that he lost friends as he won arguments. Gerhardt's arguments were set out in his "Classification chimique."

152 Jean F. Persoz, *Introduction a l'étude de la chimie moléculaire* (Paris, 1839), p. 860.

"mixed ethers," in which the two hydrocarbon radicals were not identical.[153] The subtlety of Williamson's work would scarcely have been possible from within the confines of a traditional dualism.

The fact that Berzelius's approach had to be transcended during the 1840s and 1850s must not, however, be allowed to obscure one last insight, which emerged through his dialogue with the French chemists. By 1843, as already noted, Laurent and he could agree that chlorine could directly replace hydrogen and to some extent play its role. But they still disagreed violently as to why this was possible. Laurent attributed the common properties of acetic and trichloroacetic acids to an identity of atomic arrangement in the total molecule of each: As long as the *mass* of the chlorine atom did not disturb the overall equilibrium, characteristic properties would be preserved after substitution.[154] Berzelius, on the other hand, distinguished an active (C^2O^3) from a passive (C^2H^3) or (C^2Cl^3) group within the respective acids, and proceeded to ascribe their common properties to the active (C^2O^3) group that featured in them both.[155] Because the passive group – the "copula" as Berzelius christened it – was not allowed to determine the properties of the molecule, chlorine substitution within it would be without effect. In this way, Berzelius inoculated an organic acid against an attack of chlorine, without succumbing to Laurent's contention that arrangement was the all-important variable.

How successful a device was the copula? Berzelius was under no illusions concerning its vulnerability. He addressed the question himself whether the invention and profusion of copulas in the organic domain alone might not promote unnecessary divisions between the two realms of chemistry.[156] If organic compounds really did contain inert groups of elements, would that not vitiate popular attempts to infer the internal arrangement of elements from the properties and reactions of the respective compounds?[157] Was there yet sufficient evidence to settle the question whether copulas composed of carbon and chlorine could really exist?[158] And finally, what kind of bonding could be envisaged to stick the inert copula to the active group? With such difficulties in mind, Laurent was to deride the copula as the epitome of the ad hoc. "What then is a copula?" he would inquire. "A copula is an imaginary body, the pres-

153 Priesner, "Etherification," pp. 140–145; Alexander W. Williamson, *Papers on Etherification and the Constitution of Salts*, Alembic Club reprint no. 16 (Edinburgh, 1949). See also Rocke, *Atomism*, p. 215.

154 Laurent to Berzelius, May 1843, in *Bref*, Vol. III:ii, p. 184.

155 Thus acetic acid, in Berzelius's revised view, became $[(C^2H^3)]{\cdot}C^2O^3 + H^2O]$ and trichloroacetic acid became $[(C^2Cl^3)]{\cdot}C^2O^3 + H^2O]$.

156 Berzelius, *Traité*, Vol. V, 2nd ed. (Paris, 1848), p. 41.

157 Ibid., p. 42. On other limitations of that empiricist program, see Brooke, "Philosophy of Chemistry," pp. 407–415.

158 Berzelius to Laurent, 20 October 1843, in *Bref*, Vol. III:ii, p. 191.

ence of which disguises all the chemical properties of the compounds with which it is united." [159]

For a balanced assessment, however, we should not leave Berzelius at the mercy of his detractors. As imaginary bodies go, the copula had much to commend it. In the specific case of the organic acids it was not without empirical support,[160] but its real strength and longer-term significance lay in that distinction between active and passive groups, which allowed Berzelius to retrieve some importance for the electrochemical nature of the elements. Refusing to allow the individuality of the elements to be altogether suppressed, Berzelius recognized that it was an individuality that could still assert itself in the active part of the molecule. Though the term copula was eventually dropped, the essential insight to which it corresponded was to have a long future. Functional groups of elements in organic compounds do assume an active or passive role, depending on their respective conjuncts. Toluene and ethyl benzene, for example, may be said to share common aromatic properties because the benzene ring common to them both is the dominant, or active, group. To call their aliphatic side chains copulas in Berzelius's sense would scarcely be misleading even today.

Reference to a long future does, however, raise the specter of whiggishness. Can a balanced assessment ever be given without some reference to future developments? This is not such a hackneyed question in the case of Berzelius and the future of organic chemistry. It is a question that has to be faced because, with the advent of electronic theories of organic combination in the early years of the twentieth century, Berzelius's reputation was given a lift. When J. J. Thomson sought to reintroduce electrical mechanisms into organic chemistry, it was Berzelius he cited for his precedent.[161] As David Hull has pointed out, it is difficult and perhaps even undesirable, for the historian to discard all forms of "presentism" unless he shirks the task of evaluation.[162] A sympathetic understanding of Berzelius can be achieved without reference to twentieth-century developments, but it has not been possible without making certain judgments about the durability of his concepts. That being so, it is certainly tempting to invoke twentieth-century electronic theories as a way of consolidating a more favorable estimate of Berzelius's contribution. At this point it is, however, necessary to sound a note of caution, for the case for revision can be pressed too hard if it is inspired by retrospective precepts. I should like to make the point by constructing a particular form

159 Laurent, *Chemical Method*, p. 204.
160 Brooke, "Chlorine Substitution," p. 58.
161 Martin Saltzman, "J. J. Thomson and the Modern Revival of Dualism," *Journal of Chemical Education*, 1973, 50:59–61.
162 David L. Hull, "In Defense of Presentism," *History and Theory*, 1979, 18:1–15.

of revisionist argument that can be shown to collapse under the weight of its anachronism.

The argument I have in mind would run like this: It is reasonable to assert a degree of continuity between Berzelius and the twentieth century, for the idea of a polarized bond was explicit in his electrochemical theory.[163] Not only that, but some twentieth-century theorists acknowledged their debt to him.[164] Not only that, but a knowledge of modern electronic theory proves him to have been even more perceptive than one might have imagined. We now know that it is an electro*positive* chlorine atom that displaces hydrogen in the cases that were under discussion in 1840, so Berzelius was in some deeper sense correct to resist the claim that electronegative chlorine had usurped the role of electropositive hydrogen. His intuitions, if not his arguments, were sound.

If one cavils at such an argument it is partly because one may be crediting Berzelius with intuitions he never had; and it is partly because twentieth-century theorists were more likely to have discovered Berzelius through their electronic theories than to have made their discoveries through Berzelius. There is, however, an even more engaging reason for rejecting it. The hypothesis that the chlorine atom that displaced the hydrogen in organic compounds might be in an electropositive state was actually proposed during the course of the debate. It would be convenient for the ultrarevisionist case had it been proposed by Berzelius. Its author? None other than Berzelius's opponent, Laurent. If Berzelius deserves some notional credit for having rightly intuited the future, then his adversary would seem to deserve even more for having brought such a future into his present. It was Laurent, not Berzelius, who, in a speculative bid to integrate electrochemical and structural concepts, suggested that one and the same element could exhibit different electrical states according to the configuration of its ulterior components.[165]

To press the revisionist case too hard can, therefore, lead to a reductio ad absurdum in which Berzelius emerges the loser. The more modest, and I hope more contextually sensitive, account given here simply stresses the fertility of the dualistic hypothesis in organic chemistry during the period in which Berzelius defended it. It had its limitations, which the French chemists duly exposed, but it also had a rationale in the shape of a principle of method that, during the 1830s had established itself as a sine qua non of progress in the science. So, at least, it seemed to Berzelius, who stuck to his guns when confronting alternative methodologies that did not make the electrochemical dualism of inorganic chemistry the basis

163 Russell, *Valency*, p. 287.
164 Russell, "Introduction," p. [xxvii].
165 Laurent to Berzelius, 5 January 1844, in *Bref*, Vol. III:ii, pp. 199–200; Brooke, "Chlorine Substitution," pp. 60–61.

for exploring organic composition. Contrary to the stereotypes that have been imposed on his later career, he was not even dogmatic in suggesting that his method was the only one to be *tried*. There was commendable detachment in his parting shot at Laurent: "We would do well for each of us amicably to follow his own course in hopes that science will profit from both."[166] Such detachment might encourage the suspicion noted at the outset – that method discourse belongs more to the rhetoric of self-defense than to actual scientific practice. But a willingness to entertain more than one methodology need not imply the impotence of each in its impact on the science. Too sharp a distinction between substance and rhetoric can be unhelpful in reconstructing the science of one such as Berzelius, for whom the principle of analogical argument functioned in an architectonic and not merely heuristic manner. His principle of method may not have determined scientific practice, but to open the organic textbooks of the 1830s and 1840s is to behold a theme and variations unmistakably shaped by it.

166 Berzelius to Laurent, June 1844, ibid., p. 208 [my translation].

9

Berzelius as a European Traveler

CARL GUSTAF BERNHARD

Yes, indeed, he was a traveler – amused, inquisitive, and observant. And he wrote travelogues that, together with his autobiographical notes and copious correspondence, convey a picture of the man on his travels, first in England during the Napoleonic Wars and then in postwar Europe. He preferred speaking French, but he also knew English and German and he had both corresponded and contended with his European colleagues before venturing forth for the first time. That was at about midsummer 1812.

After "tempest, rough weather, sea sickness, privateers and thousands of amusing and enjoyable adventures," he went ashore at Harwich, giddy from the voyage and with the ground rocking beneath his feet. Riding in a postchaise drawn by a team of four, he bowled up to London six miles an hour, noticing as he went that "the fruits of the earth were different," that the appearance of the countryside with its stately homes was "something all of its own" and that the cattle were "gigantic." At his last stop he got tipsy on English ale before riding along an "endless street" into the sooty metropolis with its population of 1,200,000, where the Swedish chaplain's housemaid found him a room for one pound a week. Exhausted by the exoticism of this strange country, he fell asleep with the ever-present, all-pervading scent of coal smoke about him.

For many years, Berzelius, who at that time was 34 years old and had already served as president of the Royal Swedish Academy of Sciences, had been longing to travel abroad from the northern fastness. He had

Translated from the Swedish by Roger Tannci.
I have frequently allowed Berzelius to speak for himself, by means of excerpts from his letters and writings, but for the reader's convenience, I have avoided burdening the text with notes and references to these publications, intended as they are for a Swedish readership. Many of those excerpts come from the great Berzelius monograph by Henrik Söderbaum (*Levnadsteckning*) and from Berzelius's correspondence with his friend Carl Palmstedt, published for the Royal Swedish Academy of Sciences in 1979 to 1983 by Dr. Jan Trofast. [The editors have added a few notes containing citation to literature in major languages. These notes are contained in brackets.]

Berzelius had a good sense of humor. This is a drawing of himself drinking champagne, when he received the Vasa Order from the king, Karl XIV Johan (Jean Baptiste Jules Bernadotte), King of Sweden and Norway. From the archives of the Royal Swedish Academy of Sciences, used with permission.

continued the advancement of chemical research in the spirit of Wilhelm Scheele and Torbern Bergman, added to which he had made a name for himself in Europe. But he had really been planning to visit France, because that was where he had come to be most appreciated, and Claude Louis Berthollet, Parisian chemist, friend of Napoleon, count, and senator, had for many years been making alluring noises, endorsed by Bernadotte, who was now heir apparent to the throne of Sweden but still living down there. Britannia ruled the waves, but on the Continent Napoleon and his legions lined the Neman ready to march on Moscow. When Sweden joined the anti-Napoleonic camp, the Parisian journey, at the instance of Crown Prince Carl Johan, became a visit to England instead.

This situation provides some hint of the conditions attendant on Berzelius's travels. He was an established scientist, at the same time both controversial and highly appreciated. Many foreign scientists wanted to visit with him and take advantage of the opportunity to be associated with him, and he had numerous influential connections abroad. In addition, he had letters of introduction both from Crown Prince Carl Johan and from the prime minister of Sweden and, above all, he was eager to make the personal acquaintance of his European colleagues.

To keep scientific contacts alive across national boundaries in those days, scientists had to deploy their time and their energies in a completely different manner from the present age of faster postal services, telecommunications, and comfortable travel. Correspondence meant a great deal more than it does today, and Berzelius was a consummate multilingual letter writer with a very legible hand. He had many pen friends, and altogether he must have written nearly 10,000 letters. It is difficult to understand how he found the energy to spare for letter writing on top of his intensive laboratory work and scientific authorship.

Berzelius was also a keen traveler by the standards of his time, and between 1812 and his death, in 1848, he went on ten or more protracted tours of Europe, not to mention innumerable journeys in Scandinavia. All this in spite of frequent colds and attacks of headache, nausea, and rheumatic pains – his migraine lunatique – discomforts that are scarcely compatible with visiting performances in foreign laboratories, social calls, academy sessions, attendance of theater performances, lectures, and occasionally perilous journeys by sea, on foot, on horseback, or all day and night in calashes that were not always proof against rainy weather. Nowadays we travel more or less regardless of the weather. In Berzelius's time, unexpected storms and tempests could mean very rough going. The crossing from Göteborg to Harwich, for example, could take a week or more and was usually preceded by a period spent waiting for decent weather and, preferably, favorable winds. With luck you might perhaps get a cabin, but numerous travelers between Dover and Calais were forced to stand on deck in the rain. Days and nights spent in cramped carriages drawn by one, two, or four horses were as numerous as the uncomfortable inns and changes of horses. Berzelius, who was interested in the different ways of traveling, describes quite a few breakdowns and various methods of preparing wheels, axles, and springs.

Because of his ailments, his program frequently included visits to well-known spas. As a specialist in mineral and water analyses, he would then take the opportunity of gauging the mineralogical and geological structure of the locality. He tested the waters of Bath, Le Mont Dore, and Karlsbad not only chemically but by external application and consumption, except in those cases where the same water was used first for bathing and then as a health-giving beverage.

In London, Sir Joseph Banks (1743–1820), "hobbling and stunted from gout," befriended Berzelius like an elder brother, advising him to try extract of meadow saffron *(Colchicum autumnale)* for his "lunatic attacks." At Windsor he was welcomed by the aged astronomer William Herschel (1738–1822), witty and ingenious and bearing a facial resemblance to Dean Kinnmanson, back home in Östergötland in Sweden. Of German origin, and at various times a clarinetist in the Prussian army and a Bristol organist, Herschel was now world famous, partly as a result of his discovery of Uranus. Berzelius now saw the remarkable telescope on which Herschel had been working for 40 years, assisted by his sister, "whom we were not allowed to see," and who was the first person to catch sight of the celestial phenomenon that Herschel, with "indefatigable German zeal," successfully proved to be a planet. Berzelius describes how this caught the attention of the king, who gave Herschel a house and money for continuing research, and he goes on: "Herschel is not such a great mathematician as many an Englishman who has not yet been prevailed upon to watch one night for astronomy. Therefore the unusually envious scholars of this nation have not infrequently tried to console themselves for his superior reputation by finding fault with his mathematical foundations."

Another original, mercurial personality was Smithson Tennant (1761–1815), "sloppy and unkempt," who, after a journey to Sweden, always went around with a tattered, grimy map of the country in his pocket. Edward Howard, a little man with an ugly face, unhealthy high color, and discontented mien, was a Catholic bigot but meek and placid, while the famous Thomas Young (1773–1829), in Berzelius's opinion, had a "head more wide-ranging than penetrating." He had a special weakness for William Hyde Wollaston (1766–1828), "a man of . . . extremely simple manners, aiming to show every kindness, and so fair-minded and moderate in his statements that it has become a common saying that anyone disputing a point with Wollaston is wrong." Sympathy and antipathy came quickly to Berzelius, and the impressions he records from his journeys come straight from the heart. He vastly enjoys learned society. He experienced, as he puts it, "a kind of intoxication," at the Royal Society, meeting all the eminent personalities whom he had admired from a distance but never really expected to behold.

He takes an ambivalent view of Sir Humphry Davy (1778–1829), to his mind the leading light of British chemistry and a man with whom he had been involved in some controversy. Davy has "the air of a gallant manqué," but as a speaker is "all fire." Enobled and married to a rich widow, worth 4,000 pounds a year, he lives like a dandy in a house "of such opulence that I began to suppose I had gone astray and that there was another Sir Humphry besides the chemist." At dinner, butlers and footmen ran about like ants, and everything seemed to be "le suprême bontemps," taken – as is so often the case here – to the point of pedantry and methodism. But in Davy's laboratory he found the right kind of disorder, for "a tidy laboratory is the sign of a lazy chemist."

Unlike many other traveling scientists, Berzelius also took an interest in the women surrounding his friends. Lady Davy sets out to make an impression with Petrarch and Rousseau on the anteroom table and Italian operatic scores on the forte-piano. Added to which – a point in her favor – she is pro-French. She seems to be disliked by the women, who make fun of her, but Berzelius takes up the cudgels on her behalf. True, she gives the impression of languid folly, but fundamentally she is a lovable personality and apparently the target of the envy of those around her.

A lot of time in England was devoted to scientific experiments, above all with Wollaston, Tennant, and Alexander Marcet (1770–1822). The wife of the last-mentioned, Jane Haldimand Marcet (1769–1858), impressed Berzelius by having paid such close attention to her future husband's chemistry lectures during their engagement that subsequently she published a textbook on the subject, but of course anonymously, as befitted a good wife.[1] Mrs. Hatchett is as ugly as sin, but Mrs. Rehausen, wife of the envoy, is beautiful and not necessarily stupid merely for being of few words. Berzelius also samples manorial life, and when he is seen with Sir John Sebright two daughters enter, followed by six more, "all of them equally tall, slim, ugly, thin and straight and utterly sans lignes. Can there by any greater misfortune of parenthood than fathering eight ugly daughters in a country with no nunneries?"

Berzelius has no liking for the English farce, a "pursuit far beyond the bounds of pleasantry." Where dramatic art is concerned, opera is his main source of enjoyment, and in London he takes the opportunity of cultivating this interest. He prefers taking his seat punctually, contrary to the high-society habit of entering toward the end of the final act so as to see the ballet tacked onto the performance. After the mellifluous sound of Catalani in *The Marriage of Figaro* and with her "ti vola fronte in coronar di rose" still ringing in his ears, he sees no point in staying on to watch Cassagli bounding through the air like a frightened ram followed by a pair of giantesses. Or else he takes the opportunity of getting the lady at his side to give him an exhaustive description of the structure of fashion among English ladies, with its generous exposure of neck and shoulders down to "the sabre slashes," that is, the folds outside the forcibly separated breasts, destroying the dynamics of anatomy to achieve "flatness in front and crookedness behind." The technique of redistributing both stomach and bosom, to which five printed pages and two drawings are devoted, demands not only patience and suffering but also two maid servants, two high-backed chairs, and a strong towel. Back home the ladies were appalled by the description in Berzelius's letters from his

---

1 [See M. Susan Lindee, "The American Career of Jane Marcet's *Conversations on Chemistry,* 1806–1853," *Isis,* 1991, 82:8–23, especially p. 12, for the "sexual politics of her work."]

travels, which, sometimes censored, were circulated among his friends, and his future mother-in-law "thought him above such trivialities."

Everyday life went on undisturbed by the war. We are told that "the guns were fired at the Tower for the victory of Salamanca," after which there were brilliant illuminations.

"If you've seen one castle you've seen them all." In museums and castles there were pictures "both better and worse" and most of the cathedrals are "in the old Gothic fashion." Berzelius is not very amused, but there are a few exceptions – for example, Christopher Wren's library at Trinity College, Cambridge. This was "the happiest and loveliest room" he had seen, admittedly not very well off for good looks but all the better endowed with ancient tomes. But the academic garb so greatly appreciated in the college was to him the merest "Harlequinade."

It was modern England that captivated him – the England of scientific research and industrialization. Gaslighting had begun to pierce the London darkness, and he made a close study of lighting devices and gasworks, Allen's miner's lamp, and Davy's safety lamp. Berzelius was in fact the great pioneer of primary research, concerned as he was with the history of the earth's creation, the structure of matter, and the nature of living processes. Obviously, then, he took the opportunity during his travels to pick up all kinds of modern methods to be used in his researches, just as he kept a weather eye open for geological phenomena and mineral deposits. He had, after all, laid the foundations of the chemical mineral system. Moreover, he was a skillful instrument maker; during his European travels he not only studied instrument making but went to see engravers, goldsmiths, fabric waterers, and procelain gilders, and he can account for the superior gilding of Sèvres procelain compared with Berlin porcelain. He also had an astonishing interest in industry: mineral-water factories, mines, steelworks, paper mills, and lime kilns. And as expert part owner of a chemical factory near Gripsholm, he was particularly interested in the manufacture of sulfuric acid and lead paint. He admired and studied all the refinements of an English brewery where, if we are to believe him, an ordinary hand earned as much as a secretary of state in Sweden.

It was mainly during his second visit to England, in 1818 on his way to France, that he made a real study of the big hospitals and social institutions. The London hospitals received good marks, but medic though he was, Berzelius was reluctant to witness surgical operations – understandably so, because the blessings of anesthesia had yet to be invented. He was impressed by the "Bedlam madhouse" where, so he was told, "thirty out of every hundred idiots admitted leave cured." It was the Quakers who had improved conditions in the mental hospitals. He praised a large institution housing 1,700 invalids in Greenwich and supporting 33,000 veterans all over England. All of them were ex-sailors. Nelson's gun carriage, fashioned to resemble a ship of the line and standing close to the hospital, bore witness to the connection. Trafalgar, after all, had been fought only 13 years previously.

Few travelers today are likely to devote the same amount of interest to the engines, wingspans, energy consumption, and landing gear of different types of civil aircraft as Berzelius did to the properties of different types of carriage, ranging from the three-wheeled patach à volonté in France to the English stage coach that carried 13 passengers – four inside and nine on top – plus the coachman and eight pairs of special springs, the design of which he actually made a drawing of in his travelogue.

Before sailing from Dover to Calais on an overcrowded ship whose railings were insufficient to accommodate all the seasick passengers, he found time to witness and to describe in detail the "sad physiological spectacle of the hanging of two miscreants." The public consisted mainly of women, and he was amazed at the chirruping enjoyment displayed by most of them.

It was at the height of his scientific career that Berzelius spent a year (1818–1819) in France as an experimental chemist and systematic mineralogist. He had enriched our knowledge of the nature of living phenomena, determined the atomic weights of most of the elements known at that time, presented his electrochemical theory for understanding the nature of chemical compounds, launched a code of chemical symbols, and laid the foundations of the science of geochemistry. He was greatly appreciated as a debater by the notable personages of Paris during this heyday of French scientific research.

Berzelius was thoroughly at home in the Parisian research community, which is described more affectionately than its English counterpart in his travel diary. Perhaps part of the reason was the great appreciation he had won in France, added to which his French was better than his English. Besides, tempermentally he seems to have found the French more congenial. "Old man Berthollet and his old lady are the most lovable people you could ever meet"; Thenard and Gay-Lussac "lecture like angels"; and the noble, good-hearted Dulong, whom he "cooked muck" with (that is, did chemical experiments) "is a research scientist of unusual profundity." The 75-year-old Abbé Haüy, founder of crystallography, with his wisps of hair, massive rupture, weak lips, and toothless gums, is "something of a scientist without whom mineralogy would not have become a science." Berzelius has nothing but good to say of the domineering, thick-skinned, heavyweight founder of paleontology, Georges Léopold Chrétien Frédéric Dagobert Cuvier, described by an American colleague as the stateliest but one of the least prepossessing men one could meet. The director of the Royal Porcelain Works in Sèvres, Alexandre Brongniart, and his son made Berzelius almost one of their family.

The culmination of Berzelius's sojourn in Paris came with his visit to the Société d'Arcueil, a small research society founded in Arcueil by Claude Louis Berthollet (1748–1822) and the astronomer Pierre Simon de Laplace (1749–1827), "the French Newton." He traveled there at the beginning of February, living with old man Berthollet and his old lady, "cooking muck" and debating.

In view of the role played by the Société d'Arcueil in the scientific community of the time and in view of the important French connections established by Berzelius as a result of his stay there, a short description of this small private academy is called for.[2] Pierre Simon Laplace had already been Napoleon's examiner in mathematics at the École Militaire in 1785, and during the Directory Claude Louis Berthollet had taken part in Napoleon's Egyptian adventure, in the course of which he had among other things established an academy on French lines. The close personal connection between these scientists and Napoleon, connections that persisted throughout both the Consultate and the Empire, had resulted in financial resources that, for example, enabled each of them to acquire a house of his own in Arcueil, which at that time was an idyllic small village 5 kilometers away from Paris, surrounded by farmland. Here they gathered about them a number of dedicated scientists between the ages of about 20 and 40, not only for debates, lectures, and scientific reports but also for doing scientific experiments. Berthollet, to whom science meant more than money, had spent a great deal of his fortune on fitting out a laboratory in his house.

The Société d'Arcueil had a galvanizing effect on French science between 1806 and 1820, and most of its 15 members were to play an eminent part, above all in physics and chemistry, during the first half of the nineteenth century. Apart from Joseph Louis Gay-Lussac, Louis Jacques Thenard, and Pierre Louis Dulong, the society included Jean-Baptiste Biot, Étienne Louis Malus, and Siméon Denis Poisson. Another active member of the society was the polymath Alexander von Humboldt from Berlin. Berzelius was one of the first foreigners to work for any length of time at Arcueil, where together with Dulong he studied the chemical composition of water. He stayed until the end of March. When auricula, saffron, rosa semperflorens, and apricot trees began to blossom outside his window, it was time to return to Paris.

He writes fluently about life in "the great Babylon," the Paris of Balzac. He was a welcome member of the French Academy of Sciences, and

2 [See Maurice Crosland, *The Society of Arcueil: A View of French Science at the Time of Napoleon I* (London, 1967).]

he attended concerts in the home of Lavoisier's widow, where one of the woman guests is described as having "ponderous paps and a swag belly." During the Saturday ordinary at Cuvier's he meets representatives of various cultural circles – mostly fellows with long, black, ugly pantaloons. But these "comforts of the thin-shanked" are warm and a necessary accoutrement in winter, because the great tiled stoves to which people, lifting up their coattails, turn their bottoms, are "about as much use as a twig in Hell." The queen of Sweden has lingered in France, and she invites the Parisian Swedes to a banquet and ball together with her niece and "her lady-in-waiting, who actually does it for money." Summer arrives with warm weather, the leaves burgeon, and "So many Swedes are now riddled with the pox from their lustful ways, there is hardly anyone left to cast the first stone. At the moment [Hans Gabriel Trolle-]Wachtmeister and myself are the only ones – virtue's reward – with dry cocks." A laconic description of manners as good as any. Venereal diseases were a big problem. They shared a common designation and their carriers were as yet unknown. Berzelius was no innocent, and a letter to his friend Palmstedt contains a ghoulishly humorous description of sweaty exertions on the mattress of a lady of the town – and the risks of that age.

He rounded off his visit to France in 1819 with an exploratory tour of the Massif Central among the volcanic remains and remarkable lava formations of the Auvergne and Vivarais. This had been planned by the vulcanologist Pierre Louis Antoine Cordier (1777–1861), who in the current debate on the origins of basalt argued its volcanic genesis. The result was an exciting ten-day outing by carriage, on horseback, and on foot over mountains, through desolate wilderness and poverty-ridden villages. One is amazed at Berzelius's resilience in supporting the unaccustomed rigors of wet weather, cold, drafty inns and lousy beds. But he did so in order to study, at first hand, the stratification of volcanic products.[3]

While in France he had become secretary of the Royal Swedish Academy of Sciences, and in his *Annual* for 1821 he summed up his impressions as follows: "They present the geologist with the finest possible opportunity of studying the products of volcanic phenomena in every guise, and they are now the place to which all European geologists who have not yet established their concepts as to what is or is not a volcanic product make their way in order to see with their own eyes what has become so incredible to them. I too was recently fortunate enough, guided by Cordier's directions and with his work in my hand, to visit this most remarkable region, to wonder at the horrendous remains of destruction

3 [See Tore Frängsmyr, "The Geological Ideas of J. J. Berzelius," *British Journal for the History of Science*, 1976, 9:228–236.]

and to recognize the accuracy of the conclusions drawn by Cordier from his researches." Berzelius's authoritative pronouncement meant a great deal to the French scientists.

As time went on, of course, Berzelius made a name for himself as a strict judge of science, "so that," we are told, "the whole of the chemical world submitted to his judgment." Sometimes he gives the impression of awarding marks to both women and men and nations. "I am immensely pleased with them," he says of his French research colleagues, and in the same sweeping gesture he awards top marks to the entire French people – with the exception of the inane nobility at court, that is, the circle around Louis XVIII.

In Switzerland, which he visited in 1819 on his way from France to Germany, he climbed a good way up into the Alps from the Chaminoix Valley to get a close look at Mont Blanc, "radiant in the morning sunlight," after which he ventured out onto the Glacier de Bois. Out there, on the "icy ocean," Berzelius the geologist gives us a vivid description of the dynamics of both glaciers and avalanches, and down at the bottom of the valley the glacier presents "a pageant of terrible beauty." He is amazed by the boundless energy of female English tourists, whereas a number of hapless Englishmen "looked as if they had already been hanged." He himself coped astonishingly well with the rigors of the journey, despite his embonpoint, which, in the dome of St. Paul's had prevented him from getting through the opening to the uppermost lookout platform.

On his journeys, Berzelius always had his blowpipe with him for mineral analyses, and Switzerland, like England and France, gave him abundant opportunities to practice his art in private mineral collections, laboratories, and mineralogical museums. Here again, his route took him past watering places where the water itself was of varying quality. At one of the less known establishments, "which was more than just baths," the attendant said that one had to take at least one bottle of wine, "gave us two for one and had us paid court to by fair maidens." Unfortunately he had the occasional bout of migraine. Sulfated talc did not help, but faced with the choice of eight leeches on his head or an excursion to Ferney he opted for Ferney, and arriving there he found the room exactly as old Voltaire had left it.

During the years that followed, most of Berzelius's travels took him to Germany, where in time he visited all the scientific centers and innumerable spas. Many of his successful research pupils came from Germany, and on their return there, Berzelius saw to it that desirable academic appointments were found for them. Sweden was not the only place where Berzelius held sway. His visits to Tübingen, Stuttgart, Dresden, Bonn, Göttingen, Hamburg, and Berlin meant a great deal in this connection.

On one occasion he also made a detour through Holland and Belgium, and he very often traveled home by way of Denmark. He had many friends and colleagues in Copenhagen, the foremost of them being Oersted.

This time, entering Germany after his visit to Switzerland, he had to pass through numerous customs barriers at the boundaries of petty states, principalities, and towns, and he acquired firsthand experience of the "infamous German beds, with down bolsters, soft as scrambled eggs."

There is no mistaking his enthusiasm. In Tübingen, staying with Christian Gottlob Gmelin (1792–1860), who had previously collaborated with him in Stockholm, he inspected the laboratory with the same interest "as when one married sister visits the other and, with sisterly interest and curiosity, acquaints herself with the kitchen, larder, storeroom, cellar and so on."

At a watering place not far away, we are told that one could use the same bathing facility as the king if one were prepared to pay five times the price of an ordinary gentleman's bath, whereas the poor in turn had to make do with water the gentry had used already.

In Tübingen he was shocked by the "exaggeratedly barbaric, filthy appearance" of the long-haired, bearded students, which he put down to "the philosophical spirit known to us as phosphorism and in Germany going by the name of nature philosophy." To an empiricist and meticulous experimenter like Berzelius, this was just like a red rag to a bull. The sentiment is magnificently expressed in his famous letter of 1831 to his friend C. A. Agardh, with reference to the latter's botany textbook. Berzelius describes the fire of genius giving light to the researcher as "a will-o'-the-wisp in which the pen moves at the desk when trying to present anything but that which research has found, suspected, assayed and once again convinced itself of." In a letter written in 1812 to his friend Pontin he had already inveighed against the nature philosopher Schelling's "genius manufacture," describing how in Germany one could find "philosophical passions and the dallying by ignorant specialists with the truth or doctrines which they have not understood." The hateful nature philosophy was constantly incurring the boundless contempt of Berzelius, sober and empirical as he was. When, visiting Bonn during his tour of Germany in 1835, he comes to grips with the author August Wilhelm von Schlegel, his displeasure is aroused not only by this "conceited nature philosopher" but also by the entire situation: "The tone was rigid. The professor equipped with any number of grand crosses and small stars, the guests few and tedious, the tea wretched and cream mere milk." Now, in Tübingen in 1819, he looks on in bewilderment as hordes of National-Romantic students clamor for "free constitutional government" in Germany, which in his opinion presages an age that may turn out to be bloodier than the French Revolution.

Berzelius, with his keen interest in the story of the earth's creation, was a close observer of the current debate between the Plutonist and Neptunist schools – that is, the question whether the origins of soils and rocks were marine or eruptive.[4] This is what made his visit to the Freiberg mine one of the highlights of his Germany tour. The director of the mine, Abraham Werner, who was a great exponent of the Neptunist school and, according to Berzelius, "had such a great aversion to the influence of chemistry on mineralogy," had departed this life two years earlier, but his influence remained powerful. Berzelius made the perpendicular descent, on rickety ladders, into the mine, which extended 120 kilometers underground; he inspected the mineralogical laboratory, criticized the famous mining academy, but described the mineral collection as the most abundant, the loveliest, and most enlightening he had ever seen.

In Dresden, Berzelius the mineralogist revels in August II's collection of jewels, in which the size and quantity of sapphires, rubies, emeralds, and topazes surpass anything he had imagined. He also visits the art galleries, where the "best pieces" are pointed out to him, as well as "portraits of old people in which the artist has been at pains with his brush to put every wrinkle of the ageing skin onto his canvas."

The young Eilhard Mitscherlich was one of the people who in Berlin had the good fortune to meet Berzelius and describe their experiments concerning crystal geometry. As a result he began working for Berzelius in Stockholm, and later, on Berzelius's recommendation, became a professor in Berlin, where he succeeded Klapproth. In Berlin, Berzelius also rejoiced in his visit to Thomas Seebeck (1770–1831), one of Germany's leading physicists, and in demonstrations there, and he debated the atomic theory with the chancellor of the university, in whose opinion the doctrine was completely wrong. On meeting Karl Asmund Rudolphi (1771–1832), professor of anatomy, he remarked that this "is one of the few scholars who, for all their science, have not forgotten how to behave in company like other well-brought-up people."

These first journeys put new life into Berzelius's scientific activities at home, but we still have to remember that his main achievements were already an accomplished fact and that in the various places he visited he now had to defend his theses. The many fundamental contributions made by Berzelius the scientist to the knowledge of the structure and dynamics of the organic and inorganic world aroused interest and debate in a wide variety of scientific circles. He is widely known as the author of the system of chemical symbols, which came to be internationally adopted, elevating chemistry from a descriptive to a calculating science.

---

4 [For the classic account of the subject, see Archibald Geike, *The Founders of Geology*, 2nd ed. (1905; rpt., New York, 1962).]

Among medical scientists and physiologists, though, he is probably still best known for promoting knowledge of the nature of living phenomena, and geologists and mining engineers see in him above all the creator of a systematic chemistry of minerals, profoundly interested in the origins of rocks and, accordingly, in the evolution of the planet. Many of his theses provoked vehement debate. His first presentation of the chemical system of minerals, for example, "was reviewed as the most wretched absurdity discoverable" and did not come into its own until 16 years later, when, to his gratification, the British Museum applied it to its own collections and the Royal Society rewarded him with a gold Copley medal.

Many of these debates were conducted in the various countries in the scientific academies and learned societies that Berzelius was invited to visit. Their scientific quality varied, but very often these visits provided an invigorating interchange of opinions and useful personal connections. It is significant that, eventually, Berzelius was elected to membership in more than 100 learned societies in Europe. He paid repeated visits to some of the most prestigious of them, especially the Royal Society and Académie des Sciences. The friendly little Monday Club in Berlin set the pattern of his own Monday Club back home in Stockholm, and a society for "scholars, artists and businessmen" in Geneva provide a more exotic element in the rich diversity of erudite assemblies. Academic rituals, of course, vary from one place to another, and in this particular case the old boys sat in a circle on the floor, drinking tea and relating their experiences. On one occasion they might be discussing the manufacture of chisels, on another optical illusions in the botanical garden, or even the way in which a straw filled with arsenic and poked into an anthill would induce the ants to bite each other to death, or again ways of inducing a crowd of hungry rats to follow the example of cats and thus exterminating their own species.

"Excessively good appetite is my worst cause of illness," we read in a letter from Karlsbad, where Berzelius traveled in the summer of 1822 to drink the waters, perspire, walk, and undergo laxation. The visit to Germany was prompted by his gout and intermittent headaches. But his spirits were already rising during the stormy voyage from Ystad to Stralsund when, under the influence of a persistent northwesterly wind, the ladies' stomachs were "so gravely incommoded that throughout the mail packet voyage their pussies were equally indisposed." The ship had to round the island of Rügen to make an improvised landing at Perth, where the pilot's boat took the passengers close enough to shore to be carried the rest of the way. "The ladies squealed and protested for a while before allowing themselves to be seized by the strong hands of the pilots. Such is woman's nature." Berzelius had a drastic way of putting things. In a sense he was a latter-day child of the age of Bellman, Sergel, and Ehrensvärd.[5] In the words of Auguste Renoir: "People could fart in company and at the same time express themselves grammatically. Nowadays, in this century of prattlers, people do not fart in company but many of them talk like pretentious illiterates."

Berlin was an outstanding center of chemistry, thanks to Berzelius's pupil Mitscherlich. The latter proudly displayed his laboratory, "indescribably well-ordered," as well as his new home, which, Berzelius records, was "as opulent as a whore's."

The list of visitors at the Karlsbad resort comprises 1,100 names, among them princes and princesses from near and far. The Duchess of Cumberland, "one of the pleasantest women one can meet," forces Berzelius to investigate the Karlsbad water. Despite the ban on supplying samples of the water to foreign chemists, a number of specimens were bottled for analysis, but only after an eruption of Berzelian wrath at the restrictions. Taking the waters, then, restored his spirits and energies, and arrangements were then made for the historic meeting with Goethe in Eger, where Goethe, after initially ignoring Berzelius, participated in their joint blowpipe experiments. Goethe's "entire tournure," in Berzelius's opinion, was that of a well-dressed honorable old-fashioned inspector. He is more silent than talkative, expresses himself aptly but not trenchantly, and is much more of a true philosopher in his person than in his writing.

Mesmerism and magnetic circles were all the rage, and, as a medical scientist, Berzelius accepted an invitation to take part in stunts of this kind. This was because, among all his other innumerable appointments in Sweden, he was serving on a committee that had been appointed to investigate "animal magnetism," as it was called. In Paris he felt that the demonstration of tricks during hypnotic sleep was a repulsive way of playing on a pathological peculiarity of the nervous system, and in Berlin, where efforts were made to cure all manner of organic ailments with a magnetizing box in a hocus-pocus setting, Berzelius, critical as ever, would have liked to have seen "the charlatan" hanged "by strong hemped ropes like a common cheat."

Gradually Berzelius's journeys changed character. In 1828, 1830, and 1835 he visited the Continent mainly in order to attend scientific meetings in Berlin, Hamburg, and Bonn, and he became more and more a principal character of these congresses. He was not susceptible to flattery,

---

5 [Bellman is noted in note 5 to Lindroth's essay, Chapter 1. Johan Tobias Sergel (1740–1814) was a leading sculptor. Carl August Ehrensvärd (1745–1800) was a member of a noble family, a prominent naval officer of high rank, an artist, and philosopher of art.]

but he valued appreciation and was not really surprised by the distinctions that were showered upon him, taking them as a natural phenomenon.

Berzelius had quite a reputation in scientific circles as a temperamental bull in the china shop, a headstrong genius with a bent for the flamboyant. "A majestic, calm, steady northerner with the dignity of the scientific researcher" was one verdict during his visit to Germany in 1830. But internally he could present a different appearance. Due to fluctuating hypochondria and physical disorders at the least convenient moments, he was frequently apprehensive at the prospect of grand receptions – much as he liked them – and at the hopes and demands that were pinned on his participation and interest. In addition, he took the opportunity in connection with his congresses of extending his journeys to include a vast number of other places. His 1828 journey took in about 25 different places in Germany, Belgium, and Holland in the course of a month, and was of a more superficial, tourist nature than his previous visits. The Berlin meeting, which marked the conclusion, was opened by Berzelius's peer, Alexander von Humboldt – a good friend from the Arcueil days – and both his lecture and Berzelius's were attended by royalty.

The 1835 journey to France and Germany, remarkably enough, began two days after Berzelius's engagement to Betty Poppius, 24 years younger than himself, who had to stay at home. It was a 57-year-old Berzelius who returned to Paris in a wretched state of health. He actually disliked alcohol, but tincture of rhubarb and a wine diet were now prescribed, together with lukewarm vinegar baths. He "shuddered and drank," but there was no improvement. He regrets that many of his old research colleagues have been swallowed up by politics. He was stared at and feted like a master of scientific research, received by the new royalty – Louis Philippe and his court – modeled on the colossal scale by David d'Anger, and after 41 days of banquets, felt so dreadfully ill that, on the firm recommendation of the neurologist Magendie, he left Paris.

"Why," he wrote, on the way back from Paris, "am I leaving this city with such aversion? Surely nowhere else in the world has such minor scientific merit as I may have possessed been so greatly appreciated as here, nowhere else have so many people vied with one another to do me favors and to show me friendship and respect! It would be the height of churlishness not to remember this with profound gratitude!" Paris indeed showed its appreciation, as anybody can see today by catching the Métro to Port de Clichy and then walking a little way down Avenue de Clichy to Rue Berzelius.

In Bonn, where Berzelius attended the annual meeting of scientists, he felt better and, to his great pleasure, was received by his associate Friedrich Wöhler (1800–1882), who then accompanied him to Cologne and Cassel, while from Hamburg he traveled home in the company of his friend von Bonsdorff, but, as Berzelius puts it, he whined even more than I.

Berzelius paid his last visit to Germany in 1845. By then he had had disturbing attacks of loss of memory. Returning home, for example, from a visit to Crown Prince Oscar, he did not have the slightest recollection of what had happened. Neither bleeding at the neck, mustard compresses under his feet, nor the imbibing of guaiac in alcohol and decoctions of sarsaparilla were any help, and so his friend, the physician Magnus Retzius advised him to travel to Karlsbad – a perpetual last resort. He grumbled and worried about the expense, because, as he says, "my wife has been so accustomed from childhood to the assistance of her sisters and a chambermaid in dressing that she cannot possibly manage without, both she and I need the assistance of a servant, so that we must need to travel as a party of four."

The main reason given by the doctors in favor of the journey was that it would take Berzelius away from his work, and on June 12, he sailed from Ystad on the *Svithiod,* together with his wife Betty and her aunt, Mrs. Edholm, who was also poorly. In Berlin his associates Magnus, Mitscherlich, and Rose and his friend the Swedish Minister d'Ohsson received him, and this marked the beginning of an unremitting whirl of socialization. A gala banquet was given for him at the Prussian Academy of Sciences, and he trembled at the prospect of his address of thanks but, as he said: "I managed, loudly and faultlessly, to deliver the piece I had rehearsed, thereby eliciting cheers and applause. . . . I was chock-full of seltzer and renown." He received many courtesies, but all the well-intended exaggerations "could only be swallowed if one had drunk a great deal of wine, which I had not." The king, Friedrich Wilhelm IV, gave a banquet at Sanssouci, and Berzelius was amazed by the sensible austerity of the menu, which included salt herring and jacket potatoes.

Railways had now been built in Germany, and train travel was a great experience. At a speed of about 12 miles an hour and with frequent stops to replenish the water in the boiler, the journey from Berlin to Leipzig took eight hours, during which time conversation was made difficult by the clanking of the engine and the screech of the wheels on the rails. There were official tributes in Leipzig, but in Dresden Berzelius evaded such attention and was able to indulge in a visit to the opera. Although he received a good many official visits during his stay in Karlsbad, Berzelius now felt better and was looking forward to traveling to Bonn by way of Bamburg, Frankfurt, and Wiesbaden. He traveled down the Rhine to Bonn with Jenny Lind, who was anxious to remain anonymous, and

Berzelius in 1845. From the archives of the Royal Swedish Academy of Sciences, used with permission.

in Bonn itself there was a Beethoven festival with a concert by Liszt. Berzelius appreciated Jenny Lind's singing, but "I wouldn't pay a tuppence to hear Liszt again."

Göttingen, of course, was included in the itinerary, and here there were more student tributes, banquets, and official receptions. Traveling with her husband, Betty experienced what was virtually a triumphal progress. Berzelius for his part was gratified, among other things, that his friend Alexander von Humboldt was "enough of a courtier" to converse for a long time with Madame La Baronne. (Just before Berzelius married, King Karl Johan of Sweden had made him a baron.) Many times during the journey the baroness must have had cause to worry about her husband's infirmity, but we are not told much on this point. They boarded the steamer *Svithiod* once again in Travemünde on August 31, and when he finally arrived back in Stockholm, Berzelius felt that he was at all events better than when the journey began, but, he explains, "thank God I'm home again."

Despite his intermittently wretched condition during the final years of his life, Berzelius continued his work and, when the time came for a meeting of scientists in Copenhagen during the summer of 1847, his doctor friend Retzius felt that he should take part. Berzelius protested but allowed himself to be persuaded. Arriving in Denmark, he suffered attacks of ague, and during the lectures he had to resort more often than usual to his snuffbox in order to stay awake. The visit to Copenhagen was packed with festivities and tributes from morning to evening. The king and queen gave a banquet at Sorgenfri, but afterward, when they were to take a walk in the park, Berzelius was unable to accompany them and had to go back to Oersted's, where he was staying. The queen sent to his lodgings a cone with a rose in it. "I almost think the corners of my eyes watered when I received it." Berzelius was scarcely able to attend the conclusion of the meeting, when he was invested by the king of Denmark with the Grand Cross of the Dannebrog Order. This was his last visit abroad, and he had only a year to live.

Allow me, though, to round off this story in Berzelius's own words, describing the peace he experienced at his friend Trolle-Wachtmeister's residence of Årup, on his way home from Copenhagen to Stockholm:

> The company . . . stays together . . . in the shade of the lime trees and surrounded by the rose trees near them. The ladies sit round a table at their needlework, the gentlemen round about them. The master's unusually lovable humanity unites everything around him into a patriarchal family life, with a joyful harmony between all the members, to such a degree that the peacocks come to our chairs and stand there with supplicatory looks, watching for something to be proffered to them, as indeed occasionally happens. The doves fly down from the roof and pass between us or fly up onto our arms, scrutinizing our hands for a crumb of bread that may be lying there in readiness for them. So everything here is redolent of love, trust and devotion right down to the smallest circumstances.

SECONDARY LITERATURE ON BERZELIUS
IN MAJOR LANGUAGES:
A SELECTED BIBLIOGRAPHY

Bernhard, Carl Gustaf, *Through France wtih Berzelius: Live Scholars and Dead Volcanoes* (Oxford: Pergamon Press, 1989).

Brooke, John H., "Chlorine Substitution and the Future of Organic Chemistry: Methodological Issues in the Laurent-Berzelius Correspondence," *Studies in the History and Philosophy of Science*, 1973, 4:47–94.

Frängsmyr, Tore, "The Geological Ideas of J. J. Berzelius," *British Journal for the History of Science*, 1976, 9:228–236.

Holmberg, Arne, *Bibliografi över J. J. Berzelius* [Bibliography on J. J. Berzelius]. Vol. I (*Tryckta arbeten av och om Berzelius* [Printed works by and about Berzelius], Stockholm and Uppsala, 1933; Supplement i, Stockholm and Uppsala, 1936; Supplements ii and iii, Stockholm, 1953–1967). Vol. II (*Manuskript* [Manuscripts], Stockholm, 1936; Supplement, Stockholm, 1953).

Jørgensen, Bent Søren, "Berzelius und die Lebenskraft," *Centaurus*, 1964, 10:258–281.

"More on Berzelius and the Vital Force," *Journal of Chemical Education*, 1965, 42:394–396.

Jorpes, J. Erik, *Berzelius: His Life and Work* (Stockholm, 1966; Berkeley, 1970).

Melhado, Evan M., *Jacob Berzelius: The Emergence of His Chemical System* (Stockholm: Almqvist & Wiksell International; Madison: University of Wisconsin Press, 1981).

"Mineralogy and the Autonomy of Chemistry around 1800," *Lychnos*, 1990:229–262.

"Mitscherlich's Discovery of Isomorphism," *Historical Studies in the Physical Sciences*, 1980, 11:87–123.

Meyer, Ernst von, *Geschichte der Chemie von den ältesten Zeiten bis zur Gegenwart* (Leipzig, 1899), pp. 165–230; *A History of Chemistry from Earliest Times to the Present Day*, tr. George McGowan (London and New York, 1891), pp. 191–269.

Partington, J. R., *A History of Chemistry*, Vol. IV (London: Macmillan, 1964), Ch. 5 et passim.

Prandtl, Wilhelm, *Humphry Davy, Jöns Jacob Berzelius: Zwei führende Chemiker aus der ersten Hälfte des 19. Jahrhunderts*. Grosse Naturforscher, ed. H. W. Frickhinger, Vol. III (Stuttgart, 1948).

Rocke, Alan J., *Chemical Atomism in the Nineteenth Century* (Columbus: Ohio State University, 1984), pp. 66–78, Ch. 6.

Russell, Colin A., "Berzelius and the Development of the Atomic Theory," in *John Dalton and the Progress of Science,* Donald S. L. Cardwell, ed. (Manchester: Manchester University Press, 1968), pp. 259–273.

"The Electrochemical Theory of Berzelius," *Annals of Science,* 1963, *19*:117–145.

"Introduction to the Reprint Edition [of Berzelius's *Essai sur la théorie des proportions chimiques*]" (New York and London: Johnson, 1972), pp. [v]–[xlix].

Schütt, Hans-Werner, "Beudant, Berzelius und die mineralogische Spezies," *Gesnerus,* 1984, *41*:257–268.

Söderbaum, H. G., "Berzelius und Hwasser, ein Blatt aus der Geschichte der schwedischen Naturforschung," in J. Ruska, ed., *Studien zur Geschichte der Chemie* (Berlin: Springer, 1927).

Berzlius' Werden und Wachsen, 1779–1821. Monographien aus der Geschichte der Chemie, Georg W. A. Kahlbaum, ed., Vol. III (Leipzig, 1899).

INDEX